D0191483

On Aggression

By the same author

King Solomon's Ring

Man Meets Dog

Studies in Animal and Human Behavior

Civilized Man's Eight Deadly Sins

Konrad Lorenz

On Aggression

Translated by Marjorie Kerr Wilson

A Harvest/HBJ Book

A Helen and Kurt Wolff Book

Harcourt Brace Jovanovich, Publishers

New York and London

Printed in the United States of America

Library of Congress Cataloging in Publication Data

Lorenz, Konrad.
On aggression.

(A Harvest book, HB 291)
"A Helen and Kurt Wolff book."
Translation of Das sogenannte Böse.
Bibliography: p.
1. Aggressive behavior in animals. 2. Aggressiveness (Psychology) I. Title.
[QL758.5.L6713 1974] 591.5 74-5306
ISBN 0-15-668741-0

CDEFGHIJ

To My Wife

Contents

Introduction

A friend of mine, who like a true friend had taken upon himself the task of reading through the manuscript of this book critically, wrote to me, when he was already more than halfway through it: "This is the second chapter I have read with keen interest but a mounting feeling of uncertainty. Why? Because I cannot see its exact connection with the book as a whole. You must make this easier for me." His criticism was no doubt fully justified, and the purpose of this introduction is to make clear to the reader from the start the direction taken by the book as a whole and the way in which the individual chapters are related to its ultimate object.

The subject of this book is *aggression,* that is to say the fighting instinct in beast and man which is directed *against* members of the same species. The decision to write it came about through a chance combination of two circumstances. I was in the United States, first in order to give some lectures to psychiatrists, psychoanalysts, and psychologists about some comparable behavioral theories and behavioral physiology and secondly to verify through field observation on the coral reefs of Florida a hypothesis I had formed, on the basis of aquarium observations, about the aggressive behavior of certain fish and the function of their coloring in the preservation of the species. It was at the clinical hospitals that for the first time in my life I fell into conversation with psychoanalysts who did not treat the theories of Freud as inviolable dogmas but, as is appropriate in every scientific field, working hypotheses. Viewing them in this way, I came to understand much in Sigmund Freud's theories that I had previously rejected as far too audacious. Discussions of his theories of motivation revealed unexpected correspondences between the findings of psychoanalysis and behavioral physiology, which seemed all the more significant because of the differences in approach, method, and above all inductive basis between the two disciplines.

I had expected unbridgeable differences of opinion over the concept of the death wish, which, according to one of Freud's theories, is a destructive principle which exists as an opposite pole to all instincts of self-preservation. In the eyes of the behavioral scientist this hypothesis, which is foreign to biology, is not only unnecessary but false. Aggression, the effects of which are frequently equated with those of the death wish, is an instinct like any other and in natural conditions it helps just as much as any other to ensure the survival of the individual and the species. In man, whose own efforts have caused an over-rapid change in the conditions of his life, the aggressive impulse often has destructive results. But so, too, do his other instincts, if in a less dramatic way. When I expressed these views on the theory of the death wish to my psychoanalytical friends I was surprised to find myself in the position of someone trying to force a door which is already open. They pointed out to me many passages in the writings of Freud which show how little reliance he himself had placed on his dualistic hypothesis, which must have been fundamentally alien and repugnant to him as a good monist and mechanistically thinking natural scientist.

It was shortly afterwards, when I was making a field study

of coral fish in warm seas, among which the function of aggression in the preservation of the species is plain, that the impulse to write this book came to me. For behavioral science really knows so much about the natural history of aggression that it does become possible to make statements about the causes of much of its malfunctioning in man. To achieve insight into the origins of a disease is by no means the same as to discover an effective therapy, but it is certainly one of the necessary conditions for this.

I am aware that the task I have set myself makes excessive demands upon my pen. It is almost impossible to portray in words the functioning of a system in which every part is related to every other in such a way that each has a causal influence on the others. Even if one is only trying to explain a gasoline engine it is hard to know where to begin, because the person to whom one seeks to explain it can only understand the nature of the crankshaft if he has first grasped that of the connecting rods, the pistons, the valves, the camshaft, and so on. Unless one understands the elements of a complete system as a whole, one cannot understand them at all. The more complex the structure of a system is, the greater this difficulty becomes—and it must be surmounted both in one's research and in one's teaching. Unfortunately the working structure of the instinctive and culturally acquired patterns of behavior which make up the social life of man seems to be one of the most complicated systems we know on this earth. In order to make comprehensible the few causal connections which I believe I can trace right through this tangle of reciprocal effects, I must, for good or ill, go back a long way.

Fortunately the observed facts which are my starting point are fascinating in themselves. I hope that the territorial fights of the coral fish, the "quasi-moral" urges and inhibitions of social animals, the loveless married and social life of the night heron, the bloody mass battles of the brown rat, and many other remarkable behavior patterns of animals will engage the reader's interest up to the point when he reaches an understanding of the deeper connections between them.

I intend to lead him to it by following as closely as possible the route which I took myself, and this is for reasons of principle. Inductive natural science always starts without preconceptions from the observation of individual cases and proceeds from this toward the abstract law which they all obey. Most textbooks take the opposite course for the sake of brevity and clarity and set down the general before the particular. The presentation is thereby made more lucid but less convincing. It is only too easy first to evolve a theory and then to underpin it with examples, for nature is so diverse that with diligent searching one can find apparently convincing examples to support wholly abstruse hypotheses. My book would really be convincing if the reader reached the same conclusion as myself solely on the basis of the facts which I set before him. But as I cannot expect him to follow such a thorny path, let me offer in advance, by way of a signpost, a brief account of the contents of each chapter.

I start in the first two chapters with the description of simple observations of typical forms of aggressive behavior. Then in the third I proceed to the discussion of its function in the preservation of the species. In the fourth I say enough about the physiology of instinctual motivation in general and the aggressive impulse in particular to explain the spontaneity of the irresistible outbreaks which recur with rhythmical regularity. In the fifth chapter I illustrate the process of ritualization and show how the instinctive impulse newly created by it is made independent—in so far as is necessary for the later understanding of its effects in inhibiting aggression. The sixth chapter serves the same purpose: here I have tried to give a general picture of the way instinctive impulses function. In the seventh chapter concrete examples are given to show what mechanisms evolution has "invented" in order to channel aggression along harmless paths, the role played by ritual in this process, and the similarity between the patterns of behavior which arise in this way and those which in man are guided by responsible morality. These chapters give the basis for an understanding of the functioning of four very different types of social organization. The first is the anonymous crowd, which is free of all kinds of aggression but also lacks the personal awareness and cohesion of individuals. The second is the family and social life of the night heron and other birds which nest in colonies, the only structural basis of which is territorial—the defense of a given area. The third is the remarkable "large family" of rats, the members of which do not recognize one another as individuals but by the tribal smell and whose social behavior toward one another is exemplary, while they attack with bitter factional hatred every member of the species that belongs to a different tribe. The fourth type of social organization is that in which it is the bond of love and friendship between individuals which prevents the members of the society from fighting and harming one another. This form of society, the structure of which is in many ways analogous to that of men, is shown in detail by the example of the greylag goose.

After what has been said in these eleven chapters I think I can help to explain the causes of many of the ways in which aggression in man goes wrong. The twelfth chapter, "On the Virtue of Humility," should provide a further basis by disposing of certain inner obstacles which prevent many people from seeing themselves as a part of the universe and recognizing that their own behavior too obeys the laws of nature. These obstacles come first of all from rejection of the idea of causality, which is thought to contradict the fact of free will, and secondly from man's spiritual pride. The thirteenth chapter seeks to depict the present situation of mankind objectively, somewhat as a biologist from Mars might see it. In the fourteenth chapter I try to propose certain counter-measures against those malfunctions of aggression, the causes of which I believe I have identified.

On Aggression

Chapter One

Prologue in the Sea

My childhood dream of flying is realized: I am floating weightlessly in an invisible medium, gliding without effort over sunlit fields. I do not move in the way that Man, in philistine assurance of his own superiority, usually moves, with belly forward and head upward, but in the age-old manner of vertebrates with back upward and head forward. If I want to look ahead, the discomfort of bending my neck reminds me painfully that I am really an inhabitant of another world. But I seldom want to do this, for my eyes are directed downward at the things beneath me, as becomes an earthly scientist.

Peacefully, indolently, fanning with my fins, I glide over fairy-tale scenery. The setting is the coast of one of the many little islands of coral chalk, the so-called Keys, that stretch in a long chain from the south end of the Florida peninsula. The landscape is less heroic than that of a real coral reef with its wildly cleft living mountains and valleys, but just as vivid. All over the ground, which consists of ancient coral rubble, can be seen strange hemispheres of brain coral, wavy bushes of Gorgonia, and, rarely, richly branched stems of staghorn coral, while between them are variegated patches of brown, red, and gold seaweed, not to be found in the real coral reefs further out in the ocean. At intervals are loggerhead sponges, man-broad and table-high, almost appearing man-made in their ugly but symmetrical forms. No bare surfaces of lifeless stone are visible, for any space between all these organisms is filled with a thick growth of moss animals, hydroid polyps and sponges whose violet and orange-red species cover large areas; among this teeming assortment I do not even know, in some cases, whether they belong to the plant or the animal kingdom.

My effortless progress brings me gradually into shallower water where corals become fewer, but plants more numerous. Huge forests of decorative algae, shaped exactly like African acacia trees, spread themselves beneath me and create the illusion that I am floating not just man-high above Atlantic coral ground, but a hundred times higher above an Ethiopian steppe. Wide fields of turtle grass and smaller ones of eelgrass glide away beneath me, and now that there is little more than three feet of water beneath me, a glance ahead reveals a long, dark, irregular wall stretching as far as I can see to each side and completely filling the space between the illuminated sea-bed and the mirror of the surface: it is the border between sea and land, the coast of Lignum Vitae Key.

The number of fish increases rapidly; dozens shoot from under me, reminding me of photographs of Africa where herds of wild animals flee in all directions from the shadow of an airplane. In some places, above the fields of thick turtle grass, comical fat puffers remind me of partridges taking off from a cornfield, zooming up only to glide down to land again in the next field or so. Other fish, many of which have incredible but always harmonious colors, do the opposite, diving straight into the grass as I approach. A fat porcupine with lovely devil's horns over ultramarine blue eyes lies quite quietly and grins at me. I have not hurt him, but he—or one of his kind—has hurt me! A few days ago I thoughtlessly touched one of this species, the Spiny Boxfish, and the needle-sharp parrot-beak, formed by two opposing teeth, pinched me and removed a considerable piece of skin from my right forefinger. I dive down to the specimen just sighted and, using the labor-saving technique of a duck in shallow water, leaving my backside above the surface, I seize him carefully and lift him up. After several fruitless attempts to bite, he starts to take the situation seriously and blows himself up; my hand clearly feels the "cylinder strokes" of the little pump formed by the pharyngeal muscles of the fish as he sucks in water. When the elasticity of his outer skin has reached its limit and he is lying like a distended prickly ball in my hand, I let him go and am amused at the urgency with which he squirts out the pumped-in water and disappears into the seaweed.

Then I turn to the wall separating sea from land. At first glance one could imagine it to be made of volcanic tuff, so fantastically pitted is its surface and so many are the cavities which stare like the eyeholes of skulls, dark and unfathomable. In fact, the rock consists of coral skeletons, relics of the pre-ice age. One can actually see in the ancient formations the structure of coral species still extant today and, pressed between them, the shells of mussels and snails whose living counterparts still frequent these waters. We are here on *two* coral reefs: an old one which has been dead for thousands of years and a new one growing on the old, as corals, like cultures, have the habit of growing on the skeletons of their forebears.

I swim up to and along the jagged waterfront, until I find a handy, not too spiky projection which I grasp with my right hand as an anchorage. In heavenly weightlessness, cool but not cold, a stranger in a wonderland far removed from earthly cares, rocked on gentle waves, I forget myself and am all eye, a blissful breathing captive balloon!

All around me are fish, and here in the shallow water they are mostly small fish. They approach me curiously from a distance or from the hiding places to which my coming had driven them; they dart back as I clear my snorkel by blowing out the water that has condensed in it; when I breathe quietly again they come nearer, swaying up and down in time with me in the gently undulating sea. It was by watching fish that, still with a clouded vision, I first noticed certain laws of animal behavior, without at the time understanding them in the least, but ever since I have endeavored to reach this understanding.

The multiplicity of the forms surrounding me—many so near that my far-sighted eyes cannot discern them sharply—seems at first overwhelming. But after a while their individual appearances become more familiar and my gestalt perception, that most wonderful of human faculties, begins to achieve a clearer, general view of the swarms of creatures. Then I find that there are not so many species as I thought at first. Two categories of fish are at once apparent: those which come swimming in shoals, either from the open sea or along the wall, and those which, after recovering from their panic at my presence, come slowly and cautiously out of a cave or other hiding place—always singly. Of the latter I already know that even after days or weeks the same individuals are always to be found in the same dwelling. Throughout my stay at Key Largo I visited regularly, every few days, a beautiful ocellated butterfly fish in its dwelling under a capsized landing stage and I always found it at home. Among the fish wandering hither and thither in shoals are myriads of little silversides, various small herrings which live near the coast, and their untiring hunters, the needlefish, swift as arrows. Then there are gray-green snappers loitering in thousands under landing stages, breakwaters, and cliffs, and delightful blue-and-yellow-striped grunts, so called because they make a grunting noise when removed from the water. Particularly numerous and particularly lovely are the blue-striped, the white, and the yellow-striped grunts, misnomers because all three are blue-and-yellow-striped, each with a different pattern. According to my observations, all three kinds swim frequently in mixed shoals. These fish have a buccal mucous membrane of a remarkable burning-red color, only visible when, with widely opened mouth, a fish threatens a member of its own species, which naturally responds in the same manner. However, neither in the aquarium nor in the sea have I ever seen this impressive sparring lead to a serious fight.

One of the charms of these and other colorful grunts, and also of many snappers, is the fearless curiosity with which they accompany the snorkel diver. Probably they follow harmless large fish and the now almost extinct manatee, the legendary sea cow, in the same way, in the hope of catching little fish or other tiny creatures that have been scared out of cover by the large animal. The first time I swam out from my home harbor, the landing pier of Key Haven Motel in Tarvenier on Key Largo, I was deeply impressed by the enormous crowd of grunts and snappers which surrounded me so densely that it obscured my view, and which seemed to be just as strong in numbers wherever I swam. Gradually I realized that I was always escorted by exactly the same fish and that at a modest estimate there were at least a few thousand. If I swam parallel with the shore to the next pier about half a mile away, the shoal followed me for about half this distance and then suddenly turned around and raced home as fast as it could swim. When the fish under the other landing stage noticed my coming, a startling thing happened: from the darkness of the stage emerged a monster several yards high and wide, and many times this length, throwing a deep black shadow on the sunlit sea bottom as it shot toward me, and only as it drew very near did it become resolved into a crowd of friendly grunts and snappers. The first time this happened to me, I was terrified, but later on these fish became a source of reassurance rather than fear, because while they remained with me I knew that there was no large barracuda anywhere near.

Entirely different are those daring little predators, needlefish and halfbeaks, which hunt in small bands of five or six just under the surface. Their whiplike forms are almost invisible from my submarine viewpoint, for their silver flanks reflect the light in exactly the same way as the under surface of the air, more familiar to us in its Janus face as the upper surface of the water. Seen from above, they are even more difficult to discern, since they shimmer blue-green just like the water surface. In widely spread flank formation they comb the highest layers of water hunting the little silversides which frequent the water in millions, thick as snowflakes in a blizzard and gleaming like silver tinsel. These dwarfs, the silversides, are not afraid of me, for fishes of their size would be no prey for fishes of mine. I can swim through the midst of their shoals and they give way so little that sometimes I hold my breath involuntarily to avoid breathing them in, as if I were passing through an equally dense cloud of mosquitoes. The fact that I am breathing through my snorkel in another medium does not in the least inhibit this reflex. If even the smallest needlefish approaches, the little silversides dart at lightning speed in all directions, upward, downward, and even leaping above the surface, producing in a few seconds a large clear space of water, which only gradually fills up again when the predator has passed.

Although the shapes of the fat-headed grunts and snappers are so different from those of the fine, streamlined needlefish, they have one thing in common: they do not deviate too much from the usual conception of the term "fish." Among the resident cave-dwellers the situation is different: the blue angelfish, decorated in youth with yellow vertical stripes, can still be called a "normal fish," but this thing pushing its way out of a crevice between two coral blocks, weaving with hesitating backward and forward movements, this velvet-black disk with bright yellow semicircular transverse bands and a luminous ultramarine-blue border to its lower edge, is this really a fish? Or those two round little things, the size and shape of a bumblebee, hurrying by and displaying on the *rear* end a round eye bordered with blue? Or the little jewel shining from that hollow, whose body is divided by a diagonal line from the lower anterior to the upper posterior end into a deep violet-blue and a lemon-yellow half? Or this unique little piece of dark-blue starry sky, strewn with tiny pale blue lights, which in paradoxical inversion of space is emerging from a coral block *below* me? On closer examination, all these fairy-tale figures are of course perfectly ordinary fishes, not too distantly related to my old friends and collaborators, the cichlids. The starry sky, the Marine Jewel Fish, and the little fish with the blue head and back and the yellow belly and tail, called Beau Gregory by the Floridians, are in fact close relations. The orange-red bumblebee is a baby of the "Rock Beauty," and the black and yellow disk is a young black Angelfish. But what colors, and what incredible designs: one could almost imagine they were planned to create a distant effect, like a flag or a poster.

The great, rippling mirror above me; starry skies—if only tiny ones—below; swaying weightlessly in a translucent medium, surrounded by angels, lost in contemplation and awed admiration of the creation and its beauty, I thank the creator that I am still able to observe essential details: of the dull-colored fishes or the pastel-colored grunts I nearly always see several of the same species at once, swimming in close shoal formation; but of the brightly colored species within my field of vision, there is *one* blue and *one* black angelfish. Of the two baby rock beauties that have just raced by, one is in furious pursuit of the other.

I continue to observe, although, in spite of the warmth of the water, my captive-balloon position is making me feel cold.

Now in the far distance—that is, only ten or twelve yards even in clear water—I see a beau gregory approaching, in search of food. The other beau, which is close to me, sees the intruder later than I do from my lookout post, and he only notices him when he is within about four yards. Then he shoots toward him furiously, whereupon the stranger, although he is a little bigger than his adversary, switches around and flees with vigorous strokes in wild zigzags, trying to avoid the ramming movements of his pursuer; these, if they met their mark, could inflict severe wounds, and indeed one of them does for I see a glinting scale flutter to the bottom like a wilted leaf. As soon as the stranger has disappeared into the dusky blue-green distance, the victor returns to his hollow, threading his way calmly through a dense shoal of young grunts who are in search of food in front of the entrance, and the absolute equanimity with which he passes through the shoal gives the impression that he is dodging stones or other inanimate obstacles. Even the little blue angelfish, not unlike himself in shape and color, rouses not the least sign of aggression.

Shortly afterward I observe a similar altercation between two black angelfish, scarcely a finger in length; but this time it is even more dramatic. The anger of the aggressor and the panicky flight of the intruder are even more apparent—though perhaps this is because my slow human eye is better able to follow the movements of the angelfish than those of the far swifter beau gregorys, whose performance is too quick for me.

I now realize that I am rather cold, and as I climb the coral wall into the warm air and golden sun of Florida, I formulate my observations in a few short sentences: the brilliant "poster-colored" fish are all local residents, and it is only these that I have seen defending a territory. Their furious attack is directed toward members of their own species only, except, of course, in the case of predatory fish in which, however, the motive of the pursuit is hunger and not real aggressiveness. Never have I seen fish of two different species attacking each other, even if both are highly aggressive by nature.

Chapter Two

Coral Fish in the Laboratory

In the previous chapter I made use of poetic license: I did not mention that I already knew from observations in the aquarium how furiously the brightly colored coral fish fight their own species, and that I had already formed an opinion on the biological meaning of these fights. I went to Florida to test this hypothesis, and if the facts disproved it I was ready to throw it overboard—or rather to spit it out through my snorkel, for one can hardly throw something overboard when one is swimming under water. It is a good morning exercise for a research scientist to discard a pet hypothesis every day before breakfast. It keeps him young.

Some years ago I began to study brightly colored reef fish in the aquarium, impelled not only by my aesthetic pleasure in their beauty but also by my flair for interesting biological problems. The first question that occurred to me was: Why are these fish so colorful? When a biologist asks "What is the aim or purpose of something?" he is not trying to plumb the depth of meaning of the universe or of this problem in particular, but he is attempting much more humbly to find out something quite simple and, in principle, open to solution. Since we have learned, through Charles Darwin, about evolution and even something about its causes, the question "What for?" has, for the biologist, a sharply circumscribed meaning. We know that it is the *function* of an organ that alters its form, in the sense of functional improvement; and when, owing to a small, in itself fortuitous hereditary change, an organ becomes a little better and more efficient, the bearer of this character, and his descendants, will set a standard with which other, less talented members of his species cannot compete; thus in the course of time those less fit to survive will disappear from the earth's surface. This ever present phenomenon is called natural selection and is one of the two great constructors of evolution. It is mutation, plus the recombination of hereditary factors in sexual reproduction, which provides the material for natural selection. Though the process of mutation had not yet been discovered in his time, and even the word had not been coined in its present connotation, Darwin, with remarkable foresight, postulated mutation as a necessity although he never used the word.

All the innumerable, complex, and expedient structures of plant and animal bodies owe their existence to the patient work performed in the course of millions of years by mutation and selection. We are even more convinced of this than Darwin was, and, as we shall soon see, with more justification. To some people it may seem disappointing that the many forms of life, whose harmonious laws evoke our awe and whose beauty delights our aesthetic senses, have originated in such a prosaic and causally determined way. But to the scientists it is a constant source of wonder that nature has created its highest works without ever violating its own laws.

Our question "What for?" can receive a meaningful answer only in cases where both constructors of evolution have been at work in the manner just described. Our question simply asks what function the organ or character under discussion performs in the interests of the survival of the species. If we ask, "What does a cat have sharp, curved claws for?" and answer simply by saying, "To catch mice with," this does not imply a profession of any mythical teleology, but the plain statement that catching mice is the function whose survival value, by the process of natural selection, has bred cats with this particular form of claw. Unless selection is at work, the question "What for?" cannot receive an answer with any real meaning. If we find, in a central European village, a population of mongrel dogs some of whom have straight tails and others curly ones, there is no point whatever in asking what they have such tails for. This random variety of forms—mostly more or less ugly—is the product of mutation working by itself, in other words, pure chance. But whenever we come upon highly regular, differentiated, and complicated structures, such as a bird's wing or the intricate mechanism of an instinctive behavior pattern, we must ask what demands of natural selection caused them to evolve, in other words, what they are for. We ask this question with assurance, in the confident hope of an intelligible answer, for we have found that we usually get one provided the questioner perseveres enough. This is not disproved by the few exceptional cases where scientific research has not yet been able to solve some of the most important of all biological problems, such as the question of what the wonderful forms and colors of mollusk shells are for, as the inadequate eye of these animals cannot see them, even when they are not—as they often are—hidden by the skin-fold of the mantle and in the darkness of the deep sea-bed.

The loud colors of coral fish call loudly for explanation. What species-preserving function could have caused their evolution? I bought the most colorful fishes I could find and, for comparison, a few less colorful and even some really drab species. Then I made an unexpected discovery: in the case of most of the really flamboyant "poster"-colored coral fish, it is quite impossible to keep more than one individual of a species in a small aquarium. If I put several members of the same species into the tank, there were vicious fights and within a short time only the strongest fish was left alive. Later, in Florida, it impressed me deeply to watch in the sea the same scene that I had always observed in my aquarium after the fatal battles: several fish, but only one of each species, each brightly colored but each flying a different flag, living peaceably together. At a small breakwater near my hotel, *one* beau gregory, *one* small black angelfish and *one* butterfly fish lived in peaceful association. Peaceful coexistence between two individuals of a "poster"-colored species occurs, in the aquarium or in the sea, only among those fishes that live in a permanent conjugal state. Such couples were observed, in the sea, among Blue Angelfish and Beau Gregory, and in the aquarium among white-and-yellow Butterfly Fish. The partners are inseparable and it is interesting to note that they are more aggressive toward members of their own species than single fish are. I shall explain the reason for this later.

In the sea, the principle "Like avoids like" is upheld without bloodshed, owing to the fact that the conquered fish flees from the territory of his conqueror who does not pursue him far; whereas in the aquarium, where there is no escape, the winner often kills the loser, or at least claims the whole container as his territory and so intimidates the weaker fish with continual attacks that they grow much more slowly than he does; and so his dominance increases till it leads to the fatal conclusion.

In order to observe how territory "owners" normally behave, one needs a container big enough for at least two territories of a size normally commanded by the species under examination. We therefore built an aquarium six feet long, holding more than two tons of water and big enough for several such territories of various species of smaller, coastal fish. In the "poster"-colored species, the young are nearly always not only more colorful and fiercer but also more firmly attached to their territories than the adults are. Since the young are small, we could observe their behavior in a comparatively limited space.

Into this aquarium my coworker Doris Zumpe and I put small fish, one to two inches in length, of the following: seven species of butterfly fish, two species of angelfish, eight species of demoiselles (the group to which the starry skies and the beau gregory belong), two species of triggerfish, three species of wrasse, one species of doctorfish, and several species of nonposter-colored, nonaggressive fish, such as trunkfish, puffers, and others. Thus there were about twenty-five species of "poster"-colored fish, with an average of four per species, more of some, only one of others, a total of roughly a hundred individuals. They settled in very well, with almost no losses; they started to flourish—and according to program, they began to fight.

Now came the chance of counting something. When the "exact" scientist can count or measure something, he experiences a pleasure which, to the outsider, is hard to understand. Admittedly we would know only a little less about intraspecific aggression if we had not counted but our results would be much less convincing if we could only say, "Brightly colored coral fishes hardly ever bite any other species than their own"; however, we, or to be more exact, Doris, counted the bites, with the following result: since there were about one hundred fish in the aquarium and each species was represented by an average of four, the chances of a fish biting one of its own species were three to ninety-six; but the proportion of bites inflicted on members of the same species to the bites given to other species was roughly eighty-five to fifteen. And even this small number of fifteen was misleading, because these bites came almost entirely from the demoiselles which in the aquarium stay in their caves all the time, invisible from without, and attack every intruder regardless of the species. In nature, they, too, ignore fishes of other species. Later on we omitted this group and obtained much more impressive figures.

A further proportion of the bites inflicted on fishes of different species came from those individuals which had no members of their own species in the container and therefore had to discharge their anger on other objects. Their choice of objects confirmed the correctness of my supposition as convincingly as did the more exact figures. For example, there was a single member of an uncertain species of butterly fish whose form and markings were so exactly intermediate between the white-and-gold and the white-and-black butterfly fish that we called him the white-gold-black, and he evidently shared our opinion of his classification for he divided his attacks almost equally between the representatives of those two species and was never seen to bite a member of the third species. The behavior of our single blue trigger (Odonus niger) was even more interesting. The zoologist who gave this fish its Latin name can only have seen it as a corpse in formalin, for the live fish is not black but luminous blue, suffused with a delicate violet and pink, particularly evident at the edges of the fins. I bought only one specimen of this fish because I realized, from the fights in the dealer's tank, that my own tank would be too small for two of these two-and-a-half-inch fishes. In the absence of a fellow member of his species, my blue triggerfish behaved peaceably for a time, administering only a few bites, significantly distributing them between two quite different species. Firstly he pursued the so-called blue devils, near relations of the blue gregory, which had the same beautiful blue color as himself; and secondly he attacked the two members of another triggerfish species, the so-called Picasso fish. As its name indicates, the markings of this fish are extraordinarily colorful and bizarre, but it resembles the blue trigger in its outward form if not in its color. After a few months, the stronger of the two Picassos had dispatched the weaker into the realm of formalin, and a strong rivalry sprang up between the survivor and the blue trigger. Doubtless the increased aggression of the latter toward the Picasso was influenced by the fact that his old enemies, the blue devils, had meanwhile changed from the bright blue of their youth to their drab, dove-gray adult dress which had a less fight-eliciting effect. Finally, the blue trigger killed the Picasso. I could quote many more such cases where, in similar experiments, only one fish survived. In cases where, as a result of pairing, two fishes behaved as one, one pair remained, as in the brown, and the white-and-gold butterfly fish. Numerous cases are also known where other animals, besides fish, in the absence of a member of their own species discharged their aggression on other objects, choosing for the purpose close relations or species with coloring similar to their own.

These aquarium observations, confirmed by my sea studies, prove the rule that fish are far more aggressive toward their own species than toward any other.

Now there are, as I have already described, a number of species which are not nearly so aggressive as the coral fish of my experiments. When one examines the aggressive and the more or less nonaggressive species, it is evident that there is a connection between coloring, aggressiveness, and sedentary territorial habits. Among the fish that I examined in the free state, extreme aggressiveness, associated with territorial behavior and concentrated on members of the same species, is found almost exclusively in those forms whose bright poster-like color patterns proclaim their species from afar. In fact, it was this extraordinary kind of coloring that aroused my curiosity and drew my attention to the existence of a problem. Fresh-water fish can also be beautifully colorful, and in this respect many of them can hold their own with marine fish, but apart from their beauty they contrast oddly with the coral fish.

The charm of the coloring of most fresh-water fish lies in its changeability: Cichlids, Labyrinth Fish, the red, green, and blue male Stickleback, the rainbow-colored Bitterling of our home waters, and many other forms well known to us through the home aquarium, illuminate their jewels only when they are glowing with love or anger. In many of these fish the degree of their emotion can be measured by their coloring, which also shows whether aggressiveness, sexual excitement or the flight urge is uppermost. Just as a rainbow disappears when a cloud covers the sun, so the beauty of the fish fades when the emotion that produced it wanes or is superseded by another conflicting emotion, such as fear, which quickly covers the fish with drab protective coloring. In other words, the colors of all these fish are a means of expression, only appearing when they are needed. Correspondingly, the young and often the females of these species have plain camouflage coloring.

The situation is different among the aggressive coral fish. By day, their glorious dress is as constant as if it had been painted on them in fast colors. It is only before going to sleep that most of them show their capacity for changing color by putting on a nightdress whose design is amazingly different from their day attire; but as long as they are awake and active, they keep their flamboyant colors at all costs, whether they are hotly pursuing a fellow member of their species or are themselves escaping in wild zigzags from a pursuer. They would no more think of lowering their flag than would an English battleship in a novel by Forester. And even in transport containers, where they are certainly not at ease, and during illness their gorgeous colors remain unchanged; even after death it is a long time before they disappear entirely.

In all typical poster-colored coral fish, not only are male and female both brightly colored but even the tiny babies show brilliant colors which, strangely enough, are often quite different from those of the adults, and sometimes even more striking. Most amazing of all: in several forms, only the babies are multicolored, for example the starry skies mentioned on page 9, and the blue devils (page 17), both of which change with sexual maturity into drab dove-gray fish with pale yellow tail fins.

The coloring of coral fish is distributed in large, sharply contrasting areas of the body. This is quite different from the color patterns not only of most fresh-water fish but of nearly all less aggressive and less territorial fish, whose charm lies in the delicacy of their designs, the harmony of their soft coloring, and the careful "attention to detail." When you see a grunt from a distance, you see an insignificant, greenish-silver fish, and only when he is right in front of you—a thing that may easily happen with these inquisitive creatures—do you notice the gold and sky-blue hieroglyphs clothing his body like an attractively designed brocade. Without any doubt these patterns are signals for the recognition of the species by its own members, but their design is such that it can be seen only at very close quarters by members of the species in the immediate vicinity. Conversely, the poster colors of the territorially aggressive coral fish are so arranged that they can be seen and recognized from the greatest possible distance, and we know only too well that recognition of their own species provokes furious aggression in these fish.

Many people, even those with an understanding of nature, think that we biologists show a strange desire for superfluous knowledge when we want to know what functions every single colored patch on an animal fulfills in the preservation of the species, and what causes could have led to its evolution. Indeed this curiosity is often attributed to materialism and a distorted sense of values. But every question that has a reasonable answer is justifiable, and the value and beauty of a natural object is in no way affected by our finding out why it is made in this, and no other way. The scientist's attitude cannot be better expressed than as William Beebe once formulated it in his quaint manner: "The isness of things is well worth studying; but it is their whyness that makes life worth living." The rainbow is no less beautiful because we have learned to understand the laws of light refraction to which it owes its existence, and the beauty and symmetry of design, color, and movement in our fishes must excite our admiration even more when we know that their purpose is preservation of the species that they adorn. We know, with tolerable certainty, the species-preserving function of the glorious war paint of coral fish: it elicits furious reactions of territorial defense in every fish of the same species—and only of the same species—when the reacting individual is in its own territory; and it proclaims fear-inspiring readiness to fight to the intruder encroaching on foreign ground. Both functions are practically identical with those of another natural phenomenon whose beauty has inspired our poets—bird song.

If we test this theory by comparing the fighting behavior of poster-colored and non-poster-colored fishes of the same genera and in the same environment, it proves itself particularly impressively when a poster-colored and plain-colored fish belong to the same genus; for example, the Sergeant Major, with its plain transverse bands, is a peaceful schooling fish, while its generic relation, the sharp-toothed Abudefduf, a gorgeous velvet-black fish with bright blue stripes on head and thorax and a yellow transverse band on its body, is about the fiercest of all the fierce territory owners that I met with during my coral fish studies. Our large aquarium proved too small for two tiny youngsters, scarcely an inch long, of this species; one claimed for itself the whole container and the other eked out its existence in the left upper front corner behind the bubbles of the air generator which hid it from the view of its disagreeable brother. Another good example is provided by comparing fish of the butterfly fish genera. The only peaceful one I know is the four-eyed butterfly, and this is the only one whose characteristic design is broken up into such small details that it can be recognized only at very close quarters.

The most remarkable thing of all is that coral fish which are poster-colored in youth and plain-colored at sexual maturity show the same correlation between coloring and aggression: as babies they are furious defenders of their territory but as adults they are far more peaceable; in some, one has the impression that they are obliged to divest themselves of their fight-eliciting colors in order to make friendly contact between the sexes possible. This certainly applies to the demoiselle group; several times I saw a brilliantly black and white species spawning in the aquarium, for this purpose changing their striking color for a monotonous dull gray, only to hoist the flag again as soon as spawning was over.

Chapter Three

What Aggression Is Good For

What is the value of all this fighting? In nature, fighting is such an ever-present process, its behavior mechanisms and weapons are so highly developed and have so obviously arisen under the selection pressure of a species-preserving function, that it is our duty to ask this Darwinian question.

The layman, misguided by sensationalism in press and film, imagines the relationship between the various "wild beasts of the jungle" to be a bloodthirsty struggle, all against all. In a widely shown film, a Bengal tiger was seen fighting with a python, and immediately afterward the python with a crocodile. With a clear conscience I can assert that such things never occur under natural conditions. What advantage would one of these animals gain from exterminating the other? Neither of them interferes with the other's vital interests.

Darwin's expression, "the struggle for existence," is sometimes erroneously interpreted as the struggle between different species. In reality, the struggle Darwin was thinking of and which drives evolution forward is the competition between near relations. What causes a species to disappear or become transformed into a different species is the profitable "invention" that falls by chance to one or a few of its members in the everlasting gamble of hereditary change. The descendants of these lucky ones gradually outstrip all others until the particular species consists only of individuals who possess the new "invention."

There are, however, fightlike contests between members of different species: at night an owl kills and eats even well-armed birds of prey, in spite of their vigorous defense, and when these birds meet the owl by day they attack it ferociously. Almost every animal capable of self-defense, from the smallest rodent upward, fights furiously when it is cornered and has no means of escape. Besides these three particular types of inter-specific fighting, there are other, less typical cases; for instance, two cave-nesting birds of different species may fight for a nesting cavity. Something must be said here about these three types of inter-specific fighting in order to explain their peculiarity and to distinguish them from the *intra*-specific aggression which is really the subject of this book.

The survival value of inter-specific fights is much more evident than that of intra-specific contests. The way in which a predatory animal and its prey influence each other's evolution is a classical example of how the selection pressure of a certain function causes corresponding adaptations. The swiftness of the hunted ungulate forces its feline pursuers to evolve enormous leaping power and sharply armed toes. Paleontological discoveries have shown impressive examples of such evolutionary competition between weapons of attack and those of defense. The teeth of grazing animals have achieved better and better grinding power, while, in their parallel evolution, nutritional plants have devised means of protecting themselves against being eaten, as by the storage of silicates and the development of hard, wooden thorns. This kind of "fight" between the eater and the eaten never goes so far that the predator causes extinction of the prey: a state of equilibrium is always established between them, endurable by both species.

The last lions would have died of hunger long before they had killed the last pair of antelopes or zebras; or, in terms of human commercialism, the whaling industry would go bankrupt before the last whales became extinct. What directly threatens the existence of an animal species is never the "eating enemy" but the competitor. In prehistoric times man took the Dingo, a primitive domestic dog, to Australia. It ran wild there, but it did not exterminate a single species of its quarry; instead, it destroyed the large marsupial beasts of prey which ate the same animals as it did itself. The large marsupial predators, the Tasmanian Devil and the Marsupial Wolf, were far superior to the Dingo in strength, but the hunting methods of these "old-fashioned," relatively stupid and slow creatures were inferior to those of the "modern" mammal. The Dingo reduced the marsupial population to such a degree that their methods no longer "paid," and today they exist only in Tasmania, where the Dingo has never penetrated.

In yet another respect the fight between predator and prey is not a fight in the real sense of the word: the stroke of the paw with which a lion kills his prey may resemble the movements that he makes when he strikes his rival, just as a shotgun and a rifle resemble each other outwardly; but the inner motives of the hunter are basically different from those of the fighter. The buffalo which the lion fells provokes his aggression as little as the appetizing turkey which I have just seen hanging in the larder provokes mine. The differences in these inner drives can clearly be seen in the expression movements of the animal: a dog about to catch a hunted rabbit has the same kind of excitedly happy expression as he has when he greets his master or awaits some longed-for treat. From many excellent photographs it can be seen that the lion, in the dramatic moment before he springs, is in no way angry. Growling, laying the ears back, and other well-known expression movements of fighting behavior are seen in predatory animals only when they are very afraid of a wildly resisting prey, and even then the expressions are only suggested.

The opposite process, the "counteroffensive" of the prey against the predator, is more nearly related to genuine aggression. Social animals in particular take every possible chance to attack the "eating enemy" that threatens their safety. This process is called "mobbing." Crows or other birds "mob" a cat or any other nocturnal predator, if they catch sight of it by day.

The survival value of this attack on the eating enemy is self-evident. Even if the attacker is small and defenseless, he may do his enemy considerable harm. All animals which hunt singly have a chance of success only if they take their prey by surprise. If a fox is followed through the wood by a loudly screaming jay, or a sparrow hawk is pursued by a flock of warning wagtails, his hunting is spoiled for the time being. Many birds will mob an owl, if they find one in the daytime, and drive it so far away that it will hunt somewhere else the next night. In some social animals such as jackdaws and many kinds of geese, the function of mobbing is particularly interesting. In jackdaws, its most important survival value is to teach the young, inexperienced birds what a dangerous eating enemy looks like, which they do not know instinctively. Among birds, this is a unique case of traditionally acquired knowledge.

Geese and ducks "know" by very selective, innate releasing mechanisms that anything furry, red-brown, long-shaped, and slinking is extremely dangerous, but nonetheless mobbing, with its intense excitement and the gathering together of geese from far and wide, has an essentially educational character as well as a survival value; anyone who did not know it already learns: foxes may be found *here!* At a time when only part of the shore of our lake was protected by a foxproof fence, the geese kept ten or fifteen yards clear of all unfenced cover likely to conceal a fox, but in the fenced-in area they penetrated fearlessly into the thickets of young fir trees. Besides this didactic function, mobbing of predators by jackdaws and geese still has the basic, original one of making the enemy's life a burden. Jackdaws actively attack their enemy, and geese apparently intimidate it with their cries, their thronging, and their fearless advance. The great Canada geese will even follow a fox over land in a close phalanx, and I have never known a fox in this situation try to catch one of his tormentors. With ears laid back and a disgusted expression on his face, he glances back over his shoulder at the trumpeting flock and trots slowly—so as not to lose face—away from them.

Among the larger, more defense-minded herbivores which, en masse, are a match for even the biggest predators, mobbing is particularly effective; according to reliable reports, zebras will molest even a leopard if they catch him on a veldt where cover is sparse. The reaction of social attack against the wolf is still so ingrained in domestic cattle and pigs that one can sometimes land oneself in danger by going through a field of cows with a nervous dog which, instead of barking at them or at least fleeing independently, seeks refuge between the legs of its owner. Once, when I was out with my bitch Stasi, I was obliged to jump into a lake and swim for safety when a herd of young cattle half encircled us and advanced threateningly; and when he was in southern Hungary during the First World War my brother spent a pleasant afternoon up a tree with his Scotch terrier under his arm, because a herd of half-wild Hungarian swine, disturbed while grazing in the wood, encircled them, and with bared tusks and unmistakable intentions began to close in on them.

Much more could be said about these effective attacks on the real or supposed enemy. In some birds and fishes, to serve this special purpose brightly colored "aposematic" or warning colors have evolved, which predators notice and associate with unpleasant experiences with the particular species. Poisonous, evil-tasting, or otherwise specially protected animals have, in many cases, "chosen" for these warning signals the combination of red, white, and black; and it is remarkable that the Common Sheldrake and the Sumatra Barb, two creatures which have nothing in common either with each other or the above-named groups, should have done the same thing. It has long been known that Common Sheldrake mob predatory animals and that they so disgust the fox with the sight of their brightly colored plumage that they can nest safely in inhabited foxholes. I bought some Sumatra Barbs because I had asked myself why these fishes looked so poisonous; in a large communal aquarium, they immediately answered my question by mobbing big Cichlids so persistently that I had to save the giant predators from the only apparently harmless dwarfs.

There is a third form of fighting behavior, and its survival value is as easily demonstrated as that of the predator's attack on its prey or the mobbing by the prey of the eating enemy. With H. Hediger, we call this third behavior pattern the *critical reaction*. The expression "fighting like a cornered rat" has become symbolic of the desperate struggle in which the fighter stakes his all, because he cannot escape and can expect no mercy. This most violent form of fighting behavior is motivated by fear, by the most intense flight impulses whose natural outlet is prevented by the fact that the danger is too near; so the animal, not daring to turn its back on it, fights with the proverbial courage of desperation. Such a contingency may also occur when, as with the cornered rat, flight is prevented by lack of space, or by strong social ties, like those which forbid an animal to desert its brood or family. The attack which a hen or goose makes on everything that goes too near her chicks or goslings can also be classified as a critical reaction. Many animals will attack desperately when surprised by an enemy at less than a certain critical distance, whereas they would have fled if they had noticed his coming from farther away. As Hediger has described, lion tamers maneuver their great beasts of prey into their positions in the arena by playing a dangerous game with the margin between flight distance and critical distance; and thousands of big game hunting stories testify to the dangerousness of large beasts of prey in dense cover. The reason is that in such circumstances the flight distance is particularly small, because the animal feels safe, imagining that it will not be noticed by a man even if he should penetrate the cover and get quite close; but if in so doing the man oversteps the animal's critical distance, a so-called hunting accident happens quickly and disastrously.

All the cases described above, in which animals of different species fight against each other, have one thing in common: every one of the fighters gains an obvious advantage by its behavior or, at least, in the interests of preserving the species it "ought to" gain one. But intra-specific aggression, aggression in the proper and narrower sense of the word, also fulfills a species-preserving function. Here, too, the Darwinian question "What for?" may and must be asked. Many people will not see the obvious justification for this question, and those accustomed to the classical psychoanalytical way of thinking will probably regard it as a frivolous attempt to vindicate the life-destroying principle or, purely and simply, evil. The average normal civilized human being witnesses aggression only when two of his fellow citizens or two of his domestic animals fight, and therefore sees only its evil effects. In addition there is the alarming progression of aggressive actions ranging from cocks fighting in the barnyard to dogs biting each other, boys thrashing each other, young men throwing beer mugs at each other's heads, and so on to bar-room brawls about politics, and finally to wars and atom bombs.

With humanity in its present cultural and technological situation, we have good reason to consider intra-specific aggression the greatest of all dangers. We shall not improve our chances of counteracting it if we accept it as something metaphysical and inevitable, but on the other hand, we shall perhaps succeed in finding remedies if we investigate the chain of its natural causation. Wherever man has achieved the power of voluntarily guiding a natural phenomenon in a certain direction, he has owed it to his understanding of the chain of causes which formed it. Physiology, the science concerned with the normal life processes and how they fulfill their species-preserving function, forms the essential foundation for pathology, the science investigating their disturbances. Let us forget for a moment that the aggression drive has become derailed under conditions of civilization, and let us inquire impartially into its natural causes. For the reasons already given, as good Darwinians we must inquire into the species-preserving function which, under natural—or rather precultural—conditions, is fulfilled by fights within the species, and which by the process of selection has caused the advanced development of intra-specific fighting behavior in so many higher animals. It is not only fishes that fight their own species: the majority of vertebrates do so too, man included.

Darwin had already raised the question of the survival value of fighting, and he has given us an enlightening answer: It is always favorable to the future of a species if the stronger of two rivals takes possession either of the territory or of the desired female. As so often, this truth of yesterday is not the untruth of today but only a special case; ecologists have recently demonstrated a much more essential function of aggression. Ecology—derived from the Greek *oikos,* the house—is the branch of biology that deals with the manifold reciprocal relations of the organism to its natural surroundings—its "household"—which of course includes all other animals and plants native to the environment. Unless the special interests of a social organization demand close aggregation of its members, it is obviously most expedient to spread the individuals of an animal species as evenly as possible over the available habitat. To use a human analogy: if, in a certain area, a larger number of doctors, builders, and mechanics want to exist, the representatives of these professions will do well to settle as far away from each other as possible.

The danger of too dense a population of an animal species settling in one part of the available biotope and exhausting all its sources of nutrition and so starving can be obviated by a mutual repulsion acting on the animals of the same species, effecting their regular spacing out, in much the same manner as electrical charges are regularly distributed all over the surface of a spherical conductor. This, in plain terms, is the most important survival value of intra-specific aggression.

Now we can understand why the sedentary coral fish in particular are so crazily colored. There are few biotopes on earth that provide so much and such varied nutrition as a coral reef. Here fish species can, in an evolutionary sense, take up very different professions: one can support itself as an "unskilled laborer," doing what any average fish can do, hunting creatures that are neither poisonous nor armor-plated nor prickly, in other words hunting all the defenseless organisms approaching the reef from the open sea, some as "plankton," others as active swimmers "intending" to settle on the reef, as millions of free-swimming larvae of all coral-dwelling organisms do. On the other hand, another fish species may specialize in eating forms of life that live on the reef itself and are therefore equipped with some sort of protective mechanism which the hunting fish must render harmless. Corals themselves provide many different kinds of nourishment for a whole series of fish species. Pointed-jawed butterfly fish get their food parasitically from corals and other stinging animals. They search continuously in the coral stems for small prey caught in the stinging tentacles of coral polyps. As soon as they see these, they produce, by fanning with their pectoral fins, a current so directly aimed at the prey that at the required point a "parting" is made between the polyps, pressing their tentacles flat on all sides and thus enabling the fish to seize the prey almost without getting its nose stung. It always gets it just a little stung and can be seen "sneezing" and shaking its nose, but, like pepper, the sting seems to act as an agreeable stimulant. My beautiful yellow and brown butterfly fishes prefer a prey, such as a piece of fish, stuck in the tentacles of a stinging sea anemone, to the same prey swimming free in the water. Other related species have developed a stronger immunity to stings and they devour the prey together with the coral animal that has caught it. Yet other species disregard the stinging capsules of coelenterates altogether, and eat coral animals, hydroid polyps, and even big, strong, stinging sea anemones, as placidly as a cow eats grass. As well as this immunity to poison, parrot fish have evolved a strong chisellike dentition and they eat whole branches of coral including their calcareous skeleton. If you dive near a grazing herd of these beautiful, rainbow-colored fish, you can hear a cracking and crunching as though a little gravel mill were at work—and this actually corresponds with the facts, for when such a fish excretes, it rains a little shower of white sand, and the observer realizes with astonishment that most of the snow-clean coral sand covering the glades of the coral forest has obviously passed through parrot fish.

Other fish, plectognaths, to which the comical puffers, trunk, and porcupine fish belong, have specialized in cracking hard-shelled mollusks, crabs, and sea urchins; and others again, such as angelfish, specialize in snatching the lovely feather crowns that certain feather worms thrust out of their hard, calcareous tubes. Their capacity for quick retraction acts as a protection against slower predators, but some angelfish have a way of sidling up and, with a lightning sideways jerk of the mouth, seizing the worm's head at a speed surpassing its capacity for withdrawal. Even in the aquarium, where they seize prey which has no such quick reactions, these fish cannot do otherwise than snap like this.

The reef offers many other "openings" for specialized fish. There are some which remove parasites from others and which are therefore left unharmed by the fiercest predators, even when they penetrate right into the mouth cavities of their hosts to perform their hygienic work. There are others which live as parasites on large fish, punching pieces from their epidermis, and among these are the oddest fish of all: they resemble the cleaner fish so closely in color, form, and movement that, under false pretenses, they can safely approach their victims.

It is essential to consider the fact that all these opportunities for special careers, known as ecological niches, are often provided by the same cubic yard of ocean water. Because of the enormous nutritional possibilities, every fish, whatever its specialty, requires only a few square yards of sea bottom for its support, so in this small area there can be as many fish as there are ecological niches, and anyone who has watched with amazement the thronging traffic on a coral reef knows that these are legion. However, every one of this crowd is determined that no other fish of his species should settle in his territory. Specialists of other "professions" harm his livelihood as little as, to use our analogy again, the practice of a doctor harms the trade of a mechanic living in the same village.

In less densely populated biotopes where the same unit of space can support three or four species only, a resident fish or bird can "afford" to drive away all living beings, even members of species that are no real threat to his existence; but if a sedentary coral fish tried to do the same thing, it would be utterly exhausted and, moreover, would never manage to keep its territory free from the swarms of noncompetitors of different "professions." It is in the occupational interests of all sedentary species that each should determine the spatial distribution that will benefit its own individuals, entirely without consideration for other species. The colorful "poster" patterns, described in Chapter One, and the fighting reactions elicited by them, have the effect that the fish of each species keep a measured distance only from nutritional competitors of the same species. This is the very simple answer to the much discussed question of the function of the colors of coral fish.

As I have already mentioned, the species-typical song of birds has a very similar survival value to that of the visual signals of fishes. From the song of a certain bird, other birds not yet in possession of a territory recognize that in this particular place a male is proclaiming territorial rights. It is remarkable that in many species the song indicates how strong and possibly how old the singer is, in other words, how much the listener has to fear him. Among several species of birds that mark their territory acoustically, there is great individual difference of sound expression, and some observers are of the opinion that, in such species, the personal visiting card is of special significance. While Heinroth interpreted the crowing of the cock with the words, "Here is a cock!" Baeumer, the most knowledgeable of all domestic-fowl experts, heard in it the far more special announcement, "Here is the cock Balthazar!"

Among mammals, which mostly "think through their noses," it is not surprising that marking of the territory by scent plays a big role. Many methods have been tried; various scent glands have been evolved, and the most remarkable ceremonies developed around the depositing of urine and feces; of these the leg-lifting of the domestic dog is the most familiar. The objection has been raised by some students of mammals that such scent marks cannot have anything to do with territorial ownership because they are found not only in socially living mammals which do not defend single territories, but also in animals that wander far and wide; but this opinion is only partly correct. First, it has been proved that dogs and other pack-living animals recognize each other by the scent of the marks, and it would at once be apparent to the members of a pack if a nonmember presumed to lift its leg in their hunting grounds. Secondly, Leyhausen and Wolf have demonstrated the very interesting possibility that the distribution of animals of a certain species over the available biotope can be effected not only by a space plan but also by a time plan. They found that, in domestic cats living free in open country, several individuals could make use of the same hunting ground without ever coming into conflict, by using it according to a definite timetable, in the same way as our Seewiesen housewives use our communal washhouse. An additional safeguard against undesirable encounters is the scent marks which these animals—the cats, not the housewives—deposit at regular intervals wherever they go. These act like railway signals whose aim is to prevent collision between two trains. A cat finding another cat's signal on its hunting path assesses its age, and if it is very fresh it hesitates, or chooses another path; if it is a few hours old it proceeds calmly on its way.

Even in the case of animals whose territory is governed by space only, the hunting ground must not be imagined as a property determined by geographical confines; it is determined by the fact that in every individual the readiness to fight is greatest in the most familiar place, that is, in the middle of its territory. In other words, the threshold value of fight-eliciting stimuli is at its lowest where the animal feels safest, that is, where its readiness to fight is least diminished by its readiness to escape. As the distance from this "headquarters" increases, the readiness to fight decreases proportionately as the surroundings become stranger and more intimidating to the animal. If one plotted the graph of this decrease the curve would not be equally steep for all directions in space. In fish, the center of whose territory is nearly always on the bottom, the decline in readiness to fight is most marked in the vertical direction because the fish is threatened by special dangers from above.

The territory which an animal apparently possesses is thus only a matter of variations in readiness to fight, depending on the place and on various local factors inhibiting the fighting urge. In nearing the center of the territory the aggressive urge increases in geometrical ratio to the decrease in distance from this center. This increase in aggression is so great that it compensates for all differences ever to be found in adult, sexually mature animals of a species. If we know the territorial centers of two conflicting animals, such as two garden redstarts or two aquarium sticklebacks, all other things being equal, we can predict, from the place of encounter, which one will win: the one that is nearer home.

When the loser flees, the inertia of reaction of both animals leads to that phenomenon which always occurs when a time lag enters into a self-regulating process—to an oscillation. The courage of the fugitive returns as he nears his own headquarters, while that of the pursuer sinks in proportion to the distance covered in enemy territory. Finally the fugitive turns and attacks the former pursuer vigorously and unexpectedly and, as was predictable, he in his turn is beaten and driven away. The whole performance is repeated several times till both fighters come to a standstill at a certain point of balance where they threaten each other without fighting.

The position, the territorial "border," is in no way marked on the ground but is determined exclusively by a balance of power and may, if this alters in the least, for instance if one fish is replete and lazy, come to lie in a new position somewhat nearer the headquarters of the lazy one. An old record of our observations on the territorial behavior of two pairs of cichlids demonstrates this oscillation of the territorial borders. Four fish of this species were put into a large tank and at once the strongest male, A, occupied the left, back, lower corner and chased the other three mercilessly around the whole tank; in other words, he claimed the whole tank as his territory. After a few days, male B took possession of a tiny space immediately below the surface in the diagonally opposite right, front, upper corner. There he bravely resisted the attacks of the first male. This occupation of an area near the surface is in a way an act of desperation for one of these fish, because it is risking great danger from aerial predators in order to hold its own against an enemy of its own species, which, as already explained, will attack less resolutely in such a locality. In other words, the owner of such a dangerous area has, as an ally, the fear which the surface inspires in its bad neighbor. During succeeding days, the space defended by B grew visibly, expanding downward until he finally took his station in the right, front, lower corner, so gaining a much more satisfactory headquarters. Now at last he had the same chances as A, whom he quickly pressed so far back that their territories divided the tank into two almost equal parts. It was interesting to see how both fishes patrolled the border continuously, maintaining a threatening attitude. Then one morning they were doing this on the extreme right of the tank, again around the original headquarters of B, who could now scarcely call a few square inches his own. I knew at once what had happened: A had paired, and since it is characteristic of all large cichlids that both partners take part in territorial defense, B was subjected to double pressure and his territory had decreased accordingly. Next day the fish were again in the middle of the tank, threatening each other across the "border," but now there were four, because B had also taken a mate, and thus the balance of power with the A family was restored. A week later I found the border far toward the left lower area, and encroaching on A's former territory. The reason for this was that the A couple had spawned and since one of the partners was busy looking after the eggs, only one at a time was able to attend to frontier defense. As soon as the B couple had also spawned, the previous equal division of space was re-established. Julian Huxley once used a good metaphor to describe this behavior: he compared the territories to air-balloons in a closed container, pressing against each other and expanding or contracting with the slightest change of pressure in each individual one. This territorial aggression, really a very simple mechanism of behavior-physiology, gives an ideal solution to the problem of the distribution of animals of any one species over the available area in such a way that it is favorable to the species as a whole. Even the weaker specimens can exist and reproduce, if only in a very small space. This has special significance in creatures which reach sexual maturity long before they are fully grown. What a peaceful issue of the "evil principle"!

In many animals the same result is achieved without aggressive behavior. Theoretically it suffices that animals of the same species "cannot bear the smell of each other" and avoid each other accordingly. To a certain extent this applies to the smell signals deposited by cats, though behind these lies a hidden threat of active aggression. There are some vertebrates which entirely lack intra-specific aggression but which nevertheless avoid their own species meticulously. Some frogs, in particular tree frogs, live solitary lives except at mating time, and they are obviously distributed very evenly over the available habitat. As American scientists have recently discovered, this distribution is effected quite simply by the fact that every frog avoids the quacking sound of his own species. This explanation, however, does not account for the distribution of the females, for these, in most frogs, are dumb.

We can safely assume that the most important function of intra-specific aggression is the even distribution of the animals of a particular species over an inhabitable area, but it is certainly not its only one. Charles Darwin had already observed that sexual selection, the selection of the best and strongest animals for reproduction, was furthered by the fighting of rival animals, particularly males. The strength of the father directly affects the welfare of the children in those species in which he plays an active part in their care and defense. The correlation between male parental care and rival fighting is clear, particularly in those animals which are not territorial in the sense which the Cichlids demonstrate but which wander more or less nomadically, as, for example, large ungulates, ground apes, and many others. In such animals, intra-specific aggression plays no essential part in the "spacing out" of the species. Bisons, antelopes, horses, etc., form large herds, and territorial borders and territorial jealousy are unknown to them since there is enough food for all. Nevertheless the males of these species fight each other violently and dramatically, and there is no doubt that the selection resulting from this aggressive behavior leads to the evolution of particularly strong and courageous defenders of family and herd; conversely, there is just as little doubt that the survival value of herd defense has resulted in selective breeding for hard rival fights. This interaction has produced impressive fighters such as bull bison or the males of the large baboon species; at every threat to the community, these valiantly surround and protect the weaker members of the herd.

In connection with rival fights attention must be drawn to a fact which, though it seems paradoxical to the nonbiologist, is, as we shall show later on in this book, of the very greatest importance: purely intra-specific selective breeding can lead to the development of forms and behavior patterns which are not only nonadaptive but can even have adverse effects on species preservation. This is why, in the last paragraph, I emphasized the fact that family defense, a form of strife with the extra-specific environment, has evolved the rival fight, and this in its turn has developed the powerful males. If sexual rivalry, or any other form of intra-specific competition, exerts selection pressure uninfluenced by any environmental exigencies, it may develop in a direction which is quite unadaptive to environment, and irrelevant, if not positively detrimental, to survival. This process may give rise to bizarre physical forms of no use to the species. The antlers of stags, for example, were developed in the service of rival fights, and a stag without them has little hope of producing progeny. Otherwise antlers are useless, for male stags defend themselves against beasts of prey with their fore-hoofs only and never with their antlers. Only the reindeer has based an invention on this necessity and "learned" to shovel snow with a widened point of its antlers.

Sexual selection by the female often has the same results as the rival fights. Wherever we find exaggerated development of colorful feathers, bizarre forms, etc., in the male, we may suspect that the males no longer fight but that the last word in the choice of a mate is spoken by the female, and that the male has no means of contesting this decision. Birds of Paradise, the Ruff, the Mandarin Duck, and the Argus Pheasant show examples of such behavior. The Argus hen pheasant reacts to the large secondary wing feathers of the cock; they are decorated with beautiful eye spots and the cock spreads them before her during courtship. They are so huge that the cock can scarcely fly, and the bigger they are the more they stimulate the hen. The number of progeny produced by a cock in a certain period of time is in direct proportion to the length of these feathers, and, even if their extreme development is unfavorable in other ways—his unwieldiness may cause him to be eaten by a predator while a rival with less absurdly exaggerated wings may escape—he will nevertheless leave more descendants than will a plainer cock. So the predisposition to huge wing feathers is preserved, quite against the interests of the species. One could well imagine an Argus hen that reacted to a small red spot on the wings of the male, which would disappear when he folded his wings and interfere neither with his flying capacity nor with his protective color, but the evolution of the Argus pheasant has run itself into a blind alley. The males continue to compete in producing the largest possible wing feathers, and these birds will never reach a sensible solution and "decide" to stop this nonsense at once.

Here for the first time we are up against a strange and almost uncanny phenomenon. We know that the techniques of trial and error used by the great master builders sometimes lead inevitably to plans that fall short of perfect efficiency. In the plant and animal worlds there are, besides the efficient, quantities of characteristics which only just avoid leading the particular species to destruction. But in the case of the Argus pheasant we have something quite different: it is not only like the strict efficiency expert "closing an eye" and letting second-rate construction pass in the interests of experiment, but it is selection itself that has here run into a blind alley which may easily result in destruction. This always happens when competition between members of a species causes selective breeding without any relation to the extra-specific environment.

My teacher, Oskar Heinroth, used to say jokingly, "Next to the wings of the Argus pheasant, the hectic life of Western civilized man is the most stupid product of intra-specific selection!" The rushed existence into which industrialized, commercialized man has precipitated himself is actually a good example of an inexpedient development caused entirely by competition between members of the same species. Human beings of today are attacked by so-called manager diseases, high blood pressure, renal atrophy, gastric ulcers, and torturing neuroses; they succumb to barbarism because they have no more time for cultural interests. And all this is unnecessary, for they could easily agree to take things more easily; theoretically they could, but in practice it is just as impossible for them as it is for the Argus pheasant to grow shorter wing feathers.

There are still worse consequences of intra-specific selection, and for obvious reasons man is particularly exposed to them: unlike any creature before him, he has mastered all hostile powers in his environment, he has exterminated the bear and the wolf and now, as the Latin proverb says, *"Homo homini lupus."* Striking support for this view comes from the work of modern American sociologists, and in his book *The Hidden Persuaders* Vance Packard gives an impressive picture of the grotesque state of affairs to which commercial competition can lead. Reading this book, one is tempted to believe that intra-specific competition is the "root of all evil" in a more direct sense than aggression can ever be.

In this chapter on the survival value of aggression, I have laid special stress on the potentially destructive effects of intra-specific selection: because of them, aggressive behavior can, more than other qualities and functions, become exaggerated to the point of the grotesque and inexpedient. In later chapters we shall see what effects it has had in several animals, for example, in the Egyptian Goose and the Brown Rat. Above all, it is more than probable that the destructive intensity of the aggression drive, still a hereditary evil of mankind, is the consequence of a process of intra-specific selection which worked on our forefathers for roughly forty thousand years, that is, throughout the Early Stone Age. When man had reached the stage of having weapons, clothing, and social organization, so overcoming the dangers of starving, freezing, and being eaten by wild animals, and these dangers ceased to be the essential factors influencing selection, an evil intra-specific selection must have set in. The factor influencing selection was now the wars waged between hostile neighboring tribes. These must have evolved in an extreme form of all those so-called "warrior virtues" which unfortunately many people still regard as desirable ideals. We shall come back to this in the last chapter of this book.

I return to the theme of the survival value of the rival fight, with the statement that this only leads to useful selection where it breeds fighters fitted for combat with extra-specific enemies as well as for intra-specific duels. The most important function of rival fighting is the selection of an aggressive family defender, and this presupposes a further function of intra-specific aggression: brood defense. This is so obvious that it requires no further comment. If it should be doubted, its truth can be demonstrated by the fact that in many animals, where only one sex cares for the brood, only that sex is really aggressive toward fellow members of the species. Among sticklebacks it is the male, in several dwarf cichlids the female. In many gallinaceous birds, only the females tend the brood, and these are often far more aggressive than the males. The same thing is said to be true of human beings.

It would be wrong to believe that the three functions of aggressive behavior dealt with in the last three chapters—namely, balanced distribution of animals of the same species over the available environment, selection of the strongest by rival fights, and defense of the young—are its only important functions in the preservation of the species. We shall see later what an indispensable part in the great complex of drives is played by aggression; it is one of those driving powers which students of behavior call "motivation"; it lies behind behavior patterns that outwardly have nothing to do with aggression, and even appear to be its very opposite. It is hard to say whether it is a paradox or a commonplace that, in the most intimate bonds between living creatures, there is a certain measure of aggression. Much more remains to be said before discussing this central problem in our natural history of aggression. The important part played by aggression in the interaction of drives within the organism is not easy to understand and still less easy to expound.

We can, however, here describe the part played by aggression in the structure of society among highly developed animals. Though many individuals interact in a social system, its inner workings are often easier to understand than the interaction of drives within the individual. A principle of organization without which a more advanced social life cannot develop in higher vertebrates is the so-called ranking order. Under this rule every individual in the society knows which one is stronger and which weaker than itself, so that everyone can retreat from the stronger and expect submission from the weaker, if they should get in each other's way. Schjelderup-Ebbe was the first to examine the ranking order in the domestic fowl and to speak of the "pecking order," an expression used to this day by writers. It seems a little odd though, to me, to speak of a pecking order even for large animals which certainly do not peck, but bite or ram. However, its wide distribution speaks for its great survival value, and therefore we must ask wherein this lies.

The most obvious answer is that it limits fighting between the members of a society, but here in contrast one may ask: Would it not have been better if aggression among members of a society were utterly inhibited? To this, a whole series of answers can be given. First, as we shall discuss very thoroughly in a later chapter (Ten, "The Bond"), the case may arise that a society, for example, a wolf pack or monkey herd, urgently needs aggression against other societies of the same species, therefore aggression should be inhibited only *inside* the horde. Secondly, a society may derive a beneficial firmness of structure from the state of tension arising inside the community from the aggression drive and its result, ranking order. In jackdaws, and in many other very social birds, ranking order leads directly to protection of the weaker ones. All social animals are "status seekers," hence there is always particularly high tension between individuals who hold immediately adjoining positions in the ranking order; conversely, this tension diminishes the further apart the two animals are in rank. Since high-ranking jackdaws, particularly males, interfere in every quarrel between two inferiors, this graduation of social tension has the desirable effect that the higher-ranking birds always intervene in favor of the losing party.

In jackdaws, another form of "authority" is already linked with the ranking position which the individual has acquired by its aggressive drive. The expression movements of a high-ranking jackdaw, particularly of an old male, are given much more attention by the colony members than those of a lower-ranking, young bird. For example, if a young bird shows fright at some meaningless stimulus, the others, especially the older ones, pay almost no attention to his expressions of fear. But if the same sort of alarm proceeds from one of the old males, all the jackdaws within sight and earshot immediately take flight. Since, in jackdaws, recognition of predatory enemies is not innate but is learned by every individual from the behavior of experienced old birds, it is probably of considerable importance that great store is set by the "opinion" of old, high-ranking, and experienced birds.

With the higher evolution of an animal species, the significance of the role played by individual experience and learning generally increases, while innate behavior, though not losing importance, becomes reduced to simpler though not less numerous elements. With this general trend in evolution, the significance attached to the experienced old animal becomes greater all the time, and it may even be said that the social coexistence of intelligent mammals has achieved a new survival value by the use it makes of the handing down of individually acquired information. Conversely, it may be said that social coexistence exerts selection pressure in the direction of better learning capacity, because in social animals this faculty benefits not only the individual but also the community. Thus longevity far beyond the age of reproductive capacity has considerable species-preserving value. We know from Fraser Darling and Margaret Altmann that in many species of deer the herd is led by an aged female, no longer hampered in her social duties by the obligations of motherhood.

All other conditions being equal, the age of an animal is, very consistently, in direct proportion to the position it holds in the ranking order of its society. It is thus advantageous if the "constructors" of behavior rely upon this consistency and if the members of the community—who cannot read the age of the experienced leader animal in its birth certificate—rate its reliability by its rank. Some time ago, collaborators of Robert M. Yerkes made the extraordinarily interesting observation that chimpanzees, animals well known to be capable of learning by imitation, copy only higher-ranking members of their species. From a group of these apes, a low-ranking individual was taken and taught to remove bananas from a specially constructed feeding apparatus by very complicated manipulations. When this ape, together with his feeding apparatus, was brought back to the group, the higher-ranking animals tried to take away the bananas which he had acquired for himself, but none of them thought of watching their inferior at work and learning something from him. Then the highest-ranking chimpanzee was removed and taught to use the apparatus in the same way, and when he was put back in the group the other members watched him with great interest and soon learned to imitate him.

S. L. Washburn and Irven de Vore observed that among free-living baboons the band was led not by a single animal but by a "senate" of several old males who maintained their superiority over the younger and physically stronger members by firmly sticking together and proving, as a united force, stronger than any single young male. In a more exactly observed case, one of the three "senators" was seen to be an almost toothless old creature while the other two were well past their prime. On one occasion when the band was in a treeless area and in danger of encountering a lion, the animals stopped and the young, strong males formed a defensive circle around the weaker animals. But the oldest male went forward alone, performed the dangerous task of finding out exactly where the lion was lying, without being seen by him, and then returned to the horde and led them, by a wide detour around the lion, to the safety of their sleeping trees. All followed him blindly, no one doubting his authority.

Let us look back on all that we have learned in this chapter from the objective observation of animals, and consider in what ways intra-specific aggression assists the preservation of an animal species. The environment is divided between the members of the species in such a way that, within the potentialities offered, everyone can exist. The best father, the best mother are chosen for the benefit of the progeny. The children are protected. The community is so organized that a few wise males, the "senate," acquire the authority essential for making and carrying out decisions for the good of the community. Though occasionally, in territorial or rival fights, by some mishap a horn may penetrate an eye or a tooth an artery, we have never found that the aim of aggression was the extermination of fellow members of the species concerned. This of course does not negate the fact that under unnatural circumstances, for example confinement, unforeseen by the "constructors" of evolution, aggressive behavior may have a destructive effect.

Let us now examine ourselves and try, without self-conceit but also without regarding ourselves as miserable sinners, to find out what we would like to do, in a state of highest violent aggressive feeling, to the person who elicited that emotion. I do not think I am claiming to be better than I am when I say that the final, drive-assuaging act, Wallace Craig's consummatory act, is not the killing of my enemy. The satisfying experience consists, in such cases, in administering a good beating, but certainly not in shooting or disemboweling; and the desired objective is not that my opponent should lie dead but that he should be soundly thrashed and humbly accept my physical and, if I am to be considered as good as a baboon, my mental superiority. And since, on principle, I only wish to thrash such fellows as deserve these humiliations, I cannot entirely condemn my instincts in this connection. However, it must be admitted that a slight deviation from nature, a coincidence that put a knife into one's hand at the critical moment, might turn an intended thrashing into manslaughter.

Summing up what has been said in this chapter, we find that aggression, far from being the diabolical, destructive principle that classical psychoanalysis makes it out to be, is really an essential part of the life-preserving organizaton of instincts. Though by accident it may function in the wrong way and cause destruction, the same is true of practically any functional part of any system. Moreover, we have not yet considered an all-important fact which we shall hear about in Chapter Ten. Mutation and selection, the great "constructors" which make genealogical trees grow upward, have chosen, of all unlikely things, the rough and spiny shoot of intra-specific aggression to bear the blossoms of personal friendship and love.

Chapter Four

The Spontaneity of Aggression

In the previous chapter, I think it has been adequately shown that the aggression of so many animals toward members of their own species is in no way detrimental to the species but, on the contrary, is essential for its preservation. However, this must not raise false hopes about the present situation of mankind. Innate behavior mechanisms can be thrown completely out of balance by small, apparently insignificant changes of environmental conditions. Inability to adapt quickly to such changes may bring about the destruction of a species, and the changes which man has wrought in his environment are by no means insignificant. An unprejudiced observer from another planet, looking upon man as he is today, in his hand the atom bomb, the product of his intelligence, in his heart the aggression drive inherited from his anthropoid ancestors, which this same intelligence cannot control, would not prophesy long life for the species. Looking at the situation as a human being whom it personally concerns, it seems like a bad dream, and it is hard to believe that aggression is anything but the pathological product of our disjointed cultural and social life.

And one could only wish it were no more than that! Knowledge of the fact that the aggression drive is a true, primarily species-preserving instinct enables us to recognize its full danger: it is the spontaneity of the instinct that makes it so dangerous. If it were merely a reaction to certain external factors, as many sociologists and psychologists maintain, the state of mankind would not be as perilous as it really is, for, in that case, the reaction-eliciting factors could be eliminated with some hope of success. It was Freud who first pointed out the essential spontaneity of instincts, though he recognized that of aggression only rather late. He also showed that lack of social contact, and above all deprivation of it (*Liebesverlust*), were among the factors strongly predisposing to facilitate aggression. However, the conclusions which many American psychologists drew from this correct surmise were erroneous. It was supposed that children would grow up less neurotic, better adapted to their social environment, and less aggressive if they were spared all disappointments and indulged in every way. An American method of education, based on these surmises, only showed that the aggressive drive, like many other instincts, springs "spontaneously" from the inner human being, and the results of this method of upbringing were countless unbearably rude children who were anything but nonaggressive. The tragic side of this tragicomedy followed when these children grew up and left home, and in place of indulgent parents were confronted with unsympathetic public opinion, for example when they entered college. American psychoanalysts have told me that, under the strain of the difficult social adaptation necessary, many such young people really became neurotic. This questionable method of education has apparently not yet died out, for a few years ago an American colleague who was working as a guest at our institute asked if he might stay on three weeks longer, not for scientific reasons, but because his wife's sister was staying with her and her three boys were "nonfrustration" children.

The completely erroneous view that animal and human behavior is predominantly reactive and that, even if it contains any innate elements at all, it can be altered, to an unlimited extent, by learning, comes from a radical misunderstanding of certain democratic principles: it is utterly at variance with these principles to admit that human beings are not born equal and that not all have equal chances of becoming ideal citizens. Moreover, for many decades the reaction, the "reflex," represented the only element of behavior which was studied by serious psychologists, while all "spontaneity" of animal behavior was left to the "vitalists," the mystically inclined observers of nature.

The fact that the central nervous system does not need to wait for stimuli, like an electric bell with a push-button, before it can respond, but that it can itself produce stimuli which give a natural, physiological explanation for the "spontaneous" behavior of animals and humans, has found recognition only in the last decades, through the work of Adrian, Paul Weiss, Kenneth Roeder, and above all Erich von Holst. The strength of the ideological prejudices involved was plainly shown by the heated and emotional debates that took place before the endogenous production of stimuli within the central nervous system became a fact generally recognized by the science of physiology.

In behavior research in its narrower sense, it was Wallace Craig who first made spontaneity the subject of scientific examination. Before him, William McDougall had opposed the words of Descartes, "*Animal non agit, agitur,*" engraved on the shield of the behaviorists, by the more correct statement, "The healthy animal is up and doing." But as a true vitalist he took this spontaneity for the result of the mystic vital force whose meaning nobody really knows. So he did not think of observing exactly the rhythmic repetition of spontaneous behavior patterns, let alone of continuously measuring the threshold values of eliciting stimuli, as his pupil Craig did later.

In a series of experiments with blond ring doves Craig removed the female from the male in a succession of gradually increasing periods. After one such period of deprivation, he experimented to see which objects were now sufficient to elicit the courtship dance of the male. A few days after the disappearance of the female of his own species, the male was ready to court a white dove which he had previously ignored. A few days later he was bowing and cooing to a stuffed pigeon, later still to a rolled-up cloth, and finally, after weeks of solitary confinement, he directed the courtship toward the empty corner of his box cage where the convergence of the straight sides offered at least an optical fixation point. Physiologically speaking, these observations mean that after a longer passivity of an instinctive behavior pattern, in this case courtship, the threshold value of its eliciting stimuli sinks. This is a widely spread and regular occurrence; Goethe expresses analogous laws in the words of Mephisto, *"Du siehst mit diesem Trank im Leibe bald Helena in jedem Weibe,"* * and—if you are a ring dove —you do so even in an old duster or in the empty corner of your cage.

In exceptional cases, the threshold lowering of eliciting stimuli can be said to sink to zero, since under certain conditions the particular instinct movement can "explode" without demonstrable external stimulation. A hand-reared starling that I owned many years ago had never in its life caught flies nor seen any other bird do so. All his life he had taken his food from a dish, filled daily. One day I saw him sitting on the head of a bronze statue in my parents' Viennese flat, and behaving most remarkably. With his head on one side, he seemed to be examining the white ceiling, then his head and eye movements gave unmistakable signs that he was following moving objects. Finally he flew off the statue and up to the ceiling, snapped at something invisible to me, returned to his

* "Having imbibed this potion, you will soon see Helena in every female."

post, and performed the prey-killing movements peculiar to all insect-eating birds. Then he swallowed, shook himself, as many birds do at the moment of inner relaxation, and settled down quietly. Dozens of times I climbed on a chair, and even carried a stepladder into the room—Viennese houses of that period have very high ceilings—to look for the prey that my starling had snatched: but not even the tiniest insect was there.

However, this increase of the readiness to react is far from being the only effect of the "damming" of an instinctive activity. If the stimuli normally releasing it fail to appear for an appreciable period, the organism as a whole is thrown into a state of general unrest and begins to search actively for the missing stimulus. In the simplest cases, this "search" consists only in an increase of random locomotion, in swimming or running around; in the most complicated it may include the highest achievements of learning and insight. Wallace Craig called this type of purposive searching "appetitive behavior." He also pointed out that literally every instinctive motor pattern, even the simplest locomotor co-ordination, gives rise to its own, autonomous appetite whenever adequate stimulation is withheld.

There are few instinctive behavior patterns in which threshold lowering and appetitive behavior are so strongly marked as they are, unfortunately, in intra-specific aggression. In the first chapter we have seen examples of threshold lowering in the butterfly fish which, in the absence of a fellow member of its own species, chose as substitute a member of the nearest related one, and in the blue triggerfish which not only attacked the nearest related triggerfish but also unrelated fish with only one eliciting factor in common with those of its own species, namely its blue coloring. In aquarium cichlids, to whose extraordinarily interesting family life we must give our further attention, a damming of the aggression which under natural conditions would be vented on hostile territorial neighbors, can very easily lead to killing of the mate. Nearly every aquarium keeper who has owned these fish has made the following almost inevitable mistake: a number of young fish of the same species are reared in a large aquarium to give them the chance of pairing in the most natural way. When this takes place, the aquarium suddenly becomes too small for the many adult fish. It contains one gloriously colored couple, happily united, and set upon driving out all the others. Since these unfortunates cannot escape, they swim around nervously in the corners near the surface, their fins tattered, or, having been frightened out of their hiding places, they race wildly around the aquarium. The humane aquarium keeper, pitying not only the hunted fish but also the couple which, having perhaps spawned in the meanwhile, is anxious about its brood, removes the fugitives and leaves the couple in sole possession of the tank. Thinking he has done his duty, he ceases to worry about the aquarium and its contents for the time being, but after a few days he sees, to his horror, that the female is floating dead on the surface, torn to ribbons, while there is nothing more to be seen of the eggs and the young.

This sad event, which takes place with predictable regularity, particularly in East Indian yellow cichlids, and in Brazilian mother-of-pearl fish, can be obviated either by leaving in the aquarium a "scapegoat," that is, a fish of the same species, or by the more humane method of using a container big enough for two pairs and dividing it in half with a glass partition, putting a pair on each side. Then each fish can discharge its healthy anger on the neighbor of the same sex—it is nearly always male against male and female against female—and neither of them thinks of attacking its own mate. It may sound funny, but we were often made aware of opacity of the partition, caused by growth of a weed, by the fact that a cichlid male was starting to be rude to his wife. As soon as the partition separating the "apartments" was cleaned, there was at once a furious but inevitably harmless clash with the neighbors, and the atmosphere was cleared inside each of the two compartments.

Analogous behavior can be observed in human beings. In the good old days when there was still a Habsburg monarchy and there were still domestic servants, I used to observe the following, regularly predictable behavior in my widowed aunt. She never kept a maid longer than eight to ten months. She was always delighted with a new servant, praised her to the skies, and swore that she had at last found the right one. In the course of the next few months her judgment cooled, she found small faults, then bigger ones, and toward the end of the stated period she discovered hateful qualities in the poor girl, who was finally discharged without a reference after a violent quarrel. After this explosion the old lady was once more prepared to find a perfect angel in her next employee.

It is not my intention to poke fun at my long-deceased and devoted aunt. I was able, or rather obliged, to observe exactly the same phenomenon in serious, self-controlled men, myself included, once when I was a prisoner of war. So-called polar disease, also known as expedition choler, attacks small groups of men who are completely dependent on one another and are thus prevented from quarreling with strangers or people outside their own circle of friends. From this it will be clear that the damming up of aggression will be the more dangerous, the better the members of the group know, understand, and like each other. In such a situation, as I know from personal experience, all aggression and intra-specific fight behavior undergo an extreme lowering of their threshold values. Subjectively this is expressed by the fact that one reacts to small mannerisms of one's best friends—such as the way in which they clear their throats or sneeze—in a way that would normally be adequate only if one had been hit by a drunkard.

Insight into the laws of this torturing phenomenon prevents

homicide but does not allay the torment. The man of perception finds an outlet by creeping out of the barracks (tent, igloo) and smashing a not too expensive object with as resounding a crash as the occasion merits. This helps a little, and is called, in the language of behavior physiology, a redirected activity (Tinbergen). As we shall hear later, this expedient is often resorted to in nature to prevent the injurious effects of aggression. But the human being without insight has been known to kill his friend.

Chapter Five

Habit, Ritual, and Magic

Redirection of the attack is evolution's most ingenious expedient for guiding aggression into harmless channels, and it is not the only one, for rarely do the great constructors, selection and mutation, rely on a *single* method. It is in the nature of their blind trial and error, or to be more exact, trial and success, that they often hit upon *several* possible ways of dealing with the same problem, and use them all to make its solution doubly and triply sure. This applies particularly to the various physiological mechanisms of behavior whose function it is to prevent the injuring and killing of members of the same species. As a prerequisite for the understanding of these mechanisms it is necessary for us to familiarize ourselves with a still mysterious, phylogenetic phenomenon laying down inviolable laws which the social behavior of many higher animals obeys much in the same way as the behavior of civilized man obeys his most sacred customs.

Shortly before the First World War when my teacher and friend, Sir Julian Huxley, was engaged in his pioneer studies on the courtship behavior of the Great Crested Grebe, he discovered the remarkable fact that certain movement patterns lose, in the course of phylogeny, their original specific function and become purely "symbolic" ceremonies. He called this process ritualization and used this term without quotation marks; in other words, he equated the cultural processes leading to the development of human rites with the phylogenetic processes giving rise to such remarkable "ceremonies" in animals. From a purely functional point of view this equation is justified, even bearing in mind the difference between the cultural and phylogenetic processes. I shall try to show how the astonishing analogies between the phylogenetic and cultural rites find their explanation in the similarity of their functions.

A good example of how a rite originates phylogenetically, how it acquires a meaning, and how this becomes altered in the course of further development, can be found by studying a certain ceremony of females of the duck species. This ceremony is called "inciting." As in many birds with a similar family life, the females of this species are smaller but no less aggressive than the males. Thus in quarrels between two couples it often happens that the duck, impelled by anger, advances too near the enemy couple, then gets "frightened by her own courage," turns around, and hurries back to her own strong, protective drake. Beside him, she gathers new courage and begins to threaten the neighbors again, without however leaving the safe proximity of her mate.

In its original form, this succession of behavior patterns is variable according to the varying force of the conflicting drives by which the duck is impelled. The successive dominance of aggression, fear, protection-seeking, and renewed aggressiveness can clearly be read in the expression movements and, above all, in the different positions of the duck. In our European Common Shelduck for example, the whole process, with the exception of a certain head movement coupled with a special vocal utterance, contains no ritually fixed component parts. The duck runs, as every bird of this species does when attacking, with long, lowered neck toward her opponent and immediately afterward with raised head back to her mate. She often takes refuge behind the drake, describing a semicircle around him so that finally, when she starts threatening again, she is standing beside him, with her head pointing straight forward toward the enemy couple. But if she is not in a particularly frightened mood when fleeing, she merely runs to her drake and stops in front of him. Now her breast faces the drake, so if she wants to threaten her enemy she must stretch her head and neck backward over her shoulder. If she happens to stand sideways before or behind the drake, she stretches her neck at right angles to her body axis. Thus the angle between the long axis of her body and her outstretched neck depends entirely upon her position in relation to that of her drake and that of the enemy; she shows no special preference for any of these positions of movement patterns (Fig. 1).

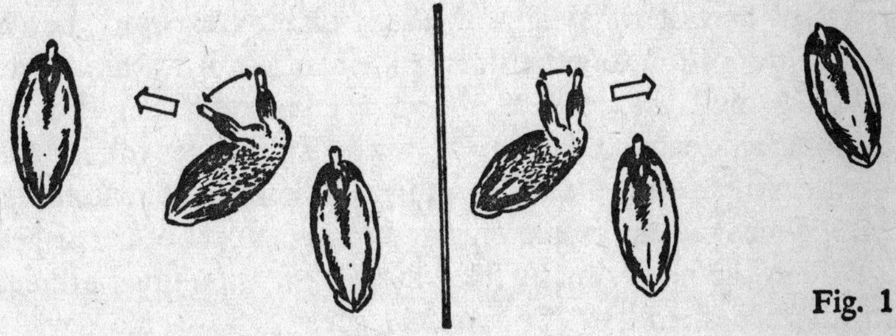

Fig. 1

In the nearly related East European-Asiatic Ruddy Sheldrake, the motor pattern of "inciting" is a small step further ritualized. In this species the duck may "still," on some occasions, stand beside her drake, threatening forward, or she may run around him, describing every kind of angle between the long axis of her body and the threatening direction of her neck; but in the majority of cases she stands with her breast to the drake, threatening backward over her shoulder. I once saw the female of an isolated couple of this species carrying out the movements of inciting without any eliciting object, and she threatened backward over her shoulder just as though she could see the nonexistent enemy in this direction.

In surface-feeding ducks, including our Mallard, the ancestor of the domestic duck, threatening backward over the shoulder has become the only possible, obligatory motor coordination. Before beginning to incite, the duck always stands with her breast as close as possible to the drake, or if he is moving she runs or swims closely after him. The head movement, directed backward over the shoulder, still contains the original orientation responses which produce, in the Ruddy Sheldrake, a motor pattern identical in its phenotype, that is, in its outer appearance, but composed of independently variable elements. This is best seen when the duck begins to perform the movement in a mild state of excitation and gradually works herself into a fury. If the enemy is standing directly in front of her, she may first threaten directly forward, but in direct proportion to her rising excitement, an irresistible force seems to pull her head backward over her shoulder. Yet an orientation reaction is still at work, striving to direct her threatening toward the enemy; this can literally be read in her eyes, which remain resolutely fixed on the object of her anger, although the new, ritually fixed movement is pulling her head in another direction. If she could speak, she would say, "I want to threaten that odious, strange drake but my head is being pulled in another direction." The existence of two conflicting directional tendencies can be demonstrated objectively: if the enemy bird is standing in front of the duck the deflection of her head backward over the shoulder is least, and it increases in direct proportion to the size of the angle between the long axis of the duck and the position of the enemy. If he is standing directly behind her, that is at an angle of 180°, she almost touches her tail with her beak (Fig. 2).

This conflict behavior observed in most female dabbling

Fig. 2

ducks has only one explanation, which must be correct however remarkable it may at first seem; in addition to those factors which originally produced the movements described, and which are easy to understand, there has evolved, in the course of phylogeny, a further, new one. In the Common Shelduck, the flight toward the drake and the attack on the enemy suffice to explain the behavior of the duck; in the Mallard, the same impulses are obviously still at work, but the behavior pattern determined by them is superseded by an independent new motor co-ordination. Analysis of the whole process is made extremely difficult by the fact that the new fixed motor pattern, which has arisen by "ritualization," is a hereditarily fixed copy of a behavior pattern originally induced by several other motives. The original behavior differs from case to case according to the varying force of each separate, independently variable impulse; the newly arisen, fixed motor co-ordination represents only one stereotyped average case. This has now become "schematized" in a manner strongly reminiscent of symbols in human cultural history. In the Mallard, the original variability of the positions in which drake and enemy may be situated is schematically programmed so that the drake must stand in front of the duck and the enemy behind her. The retreat toward the drake, originally motivated by escape drive, and the aggressive advance on the enemy are welded into one fixed, ceremonial to-and-fro movement whose rhythmical repetition increases its effectiveness as a signal. The newly arisen fixed motor pattern does not suddenly become preponderant but exists first beside the unritualized model over which it predominates only slightly. In the Ruddy Shelduck, for example, the motor co-ordinations forcing the head backward over the shoulder are seen only when the ceremony is performed *in vacuo* like the fly-catching of my starling (page 59), that is, in the absence of the enemy at which, through dominance of the original orientation mechanisms, the threatening movements would otherwise be aimed.

The above example of inciting in the Mallard is typical of most cases of phylogenetic ritualization: a new instinctive motor pattern arises whose form copies that of a behavior pattern which is variable and which is caused by several independent motivations.

For those interested in the laws of heredity and phylogenetics it may here be said that the process described above is the exact opposite of the so-called phenocopy. We speak of this when through extrinsic individually acting influences, an appearance, a phenotype, is produced which is identical with one that, in other cases, is determined by hereditary factors. In ritualization, a newly arisen hereditary disposition copies forms of behavior formerly caused phenotypically by the concurrence of very different environmental influences. We might well speak of a genocopy.

The example of inciting may further serve to illustrate the peculiarity of rite formation. In diving ducks, the inciting of the females is ritualized in a somewhat different and more complicated way: in the Crested Pochard, not only the enemy-threatening movement but also the protection-seeking movement is ritualized, that is established by a fixed motor pattern which has evolved *ad hoc*. The female Crested Pochard alternates rhythmically between a backward thrusting of her head over her shoulder and a pronounced turning of the head toward her drake, each time moving her chin up and down, a set of movements corresponding to a mimically exaggerated fleeing movement.

In the White-eye, the female advances threateningly some distance toward the enemy and then swims quickly back to her drake, making repeated chin-lifting movements which are here scarcely distinguishable from the movements of taking off. In the Golden-eye, inciting is almost entirely independent of the presence of a member of her species representing the enemy. The duck swims behind her drake and performs, in rhythmic regularity, extensive neck and head movements, alternately to the right backward and to the left backward. These would hardly be recognized as threatening movements if the phylogenetic intermediate steps were not known.

Just as the form of these movements, in the course of their progressive ritualization, has become different from those of the nonritualized prototype, so also has their meaning. The inciting of the Common Shelduck is "still" exactly like the ordinary threatening of the species and its effect on the drake is in no way different from that which, in species lacking a special inciting ceremony, the threatening of one member of a group has on another: the latter may be infected by the anger of the companion and join in the attack. In the somewhat stronger and more aggressive Ruddy Sheldrake and particularly in the Egyptian Goose, this originally mildly stimulating effect of inciting is many times stronger. In these birds, inciting really deserves its name, for the males react like fierce dogs which only await their master's signal to release their fury. In these species, the function of inciting is intimately connected with that of territorial defense. Heinroth found that the males could agree in a communal enclosure if all the females were removed.

In dabbling and in diving ducks, it is relatively seldom that the drake responds to the inciting of his duck by attacking the "enemy"; in this case the quotation marks are merited. In an unpaired Mallard, for example, inciting simply implies an invitation to pair, though *not* to mate: the precopulatory ceremony looks quite different and is called pumping. Inciting is the invitation to permanent pair formation. If the drake is inclined to accept the proposal, he lifts his chin, turns his head slightly away from the duck, and says very quickly, "Rabrab, rabrab," or, especially when he is in the water, he answers with a certain likewise ritualized ceremony: drinking and sham preening. Both these ceremonies mean that the drake Mallard is answering, "I will!" The utterance "Rabrab" contains an element of aggression; the turning away of the head with lifted chin is a typical gesture of appeasement. If he is very excited, the drake may actually make a small demonstration attack on another drake which chances to be standing near. In the second ceremony, drinking and sham preening, this never happens. Inciting on the one hand, and drinking and sham preening on the other, mutually elicit each other, and the couple can persist in them for a long time. Though drinking and sham preening have arisen from a gesture of embarrassment in whose original form aggression was present, this is no longer contained in the ritualized movement seen in dabbling ducks. In these birds, the ceremony acts as a pure appeasement gesture. In Crested Pochards and other diving ducks, I have never known the inciting of the duck to rouse the drake to serious attack.

Thus while the message of inciting in Ruddy Shelduck and Egyptian Geese could be expressed in the words "Drive him off, thrash him!," in diving ducks it simply means, "I love you." In several groups, midway between these two extremes, as for example in the Gadwall and the Widgeon, an intermediate meaning may be found: "You are my hero. I rely on you." Naturally the signal function of these symbols fluctuates even within the same species according to the situation, but the gradual phylogenetic change of meaning of the symbol has undoubtedly progressed in the direction indicated.

Many more examples of analogous processes could be given: for instance, in cichlids an ordinary swimming movement has become a means of summoning the young and, in a special case, a warning signal to the young; in the domestic fowl, the eating sound has become the enticement call of the cock and has given rise to sound expressions of sharply defined sexual meaning.

In insects, there is a certain differentiated series of ritual behavior patterns which I will discuss in more detail, not only because it illustrates even better than the above examples the parallels between the phylogenetic origin of such a ceremony and the cultural development of symbols, but also because in this unique case the "symbol" is not only a behavior pattern but it takes on a physical form which literally becomes an idol.

In several species of so-called Empid Flies (in German very appropriately called *Tanzfliegen*—Dancing Flies), closely related to the fly-eating Asalid Flies, a rite has developed as pretty as it is expedient. In this rite the male presents the female, immediately before copulation, with a slaughtered insect of suitable size. While she is engaged in eating it, he can mate her without fear of being eaten by her himself, a risk apparently threatening the suitors of fly-eating flies, particularly as the male is smaller than the female. Without any doubt, this menace exerted the selection pressure that has caused the evolution of this remarkable behavior. However, the ceremony has also been preserved in a species, the Hyperborean Empis, in which the female no longer eats flies except at her marriage feast. In a North American species, the male spins a pretty white balloon that attracts the female visually; it contains a few small insects which she eats during copulation. Similar conditions can be observed in the Southern Empid, Hilara maura, whose males spin little waving veils in which food is sometimes, but not always, interwoven. But in Hilara sartor, the Tailor Fly, found in Alpine regions and deserving more than all its relations the name of dancing fly, the males no longer catch flies but spin a lovely little veil, spanned during flight between the middle and hind legs, to which optical stimulus the female reacts. In the revised edition of Brehm's *Tierleben,* Heymons describes the collective courtship dance of these flies: "Hundreds of these little veil-carriers whirl through the air in their courtship dance, their tiny veils, about 2 mm. in size, glistening like opals in the sun."

In discussing the inciting ceremony of female ducks, I have tried to show how the origin of a new hereditary co-ordination plays an essential part in the formation of a new rite, and how in this way an autonomous and essentially fixed sequence of movements, a new instinctive motor pattern, arises. The example of the dancing flies is perhaps relevant to show us the other, equally important side of ritualization, namely the newly arising reaction with which the member of the species to whom the message is addressed answers it. In those dancing fly species in which the females are presented with a purely symbolic veil or balloon without edible contents, they obviously react to this idol just as well as or better than their ancestors did to the material gift of edible prey. And so there arises not only an instinctive movement which was not there before and which has a definite signal function in the one member of the species, the "actor," but also an innate understanding of it by the other, the "reactor." What appears to us, on superficial examination, as one ceremony, often consists of a whole number of behavior elements eliciting each other mutually.

The newly arisen motor co-ordination of the ritualized behavior pattern bears the character of an independent instinctive movement; the eliciting situation, too, which in such cases is largely determined by the answering behavior of the addressee, acquires all the properties of the drive-relieving end situation, aspired to for its own sake. In other words, the chain of actions that originally served other objective and subjective ends, becomes an end in itself as soon as it has become an autonomic rite. It would be misleading to call the ritualized movement pattern of inciting in the Mallard, or even in most diving ducks, the "expression" of love, or of affinity to the mate. The independent instinctive movement is not a by-product, not an "epiphenomenon" of the bond holding the two animals together, it is itself the bond. The constant repetition of these ceremonies which hold the pair together gives a good measure of the strength of the autonomous drive which sets them in motion. If a bird loses its mate, it loses the only object on which it can discharge this drive, and the way it seeks the lost partner bears all the characteristics of so-called appetitive behavior, that is the purposeful struggle to reach that relieving end situation wherein a dammed instinct can be assuaged.

What I have here tried to show is the inestimably important fact that by the process of phylogenetic ritualization a new and completely autonomous instinct may evolve which is, in principle, just as independent as any of the so-called "great" drives such as hunger, sex, fear, or aggression, and which—like these—has its seat in the great parliament of instincts. This again is important for our theme, because it is particularly the drives that have arisen by ritualization which are so often called upon, in this parliament, to oppose aggression, to divert it into harmless channels, and to inhibit those of its actions that are injurious to the survival of the species. In the chapter on the formation of personal bonds, we shall hear how rites arising from redirected aggression movements perform this most important function.

Those other rites, which evolve in the course of human civilization, are not hereditarily fixed but are transmitted by tradition and must be learned afresh by every individual. In spite of this difference, the parallel goes so far that it is quite justifiable to omit the quotation marks, as Huxley did. At the same time these functional analogies show what different causal mechanisms the great constructors use to achieve almost identical effects.

Among animals, symbols are not transmitted by tradition from generation to generation, and it is here, if one wishes, that one may draw the border line between "the animal" and man. In animals, individually acquired experience is sometimes transmitted by teaching and learning, from elder to younger individuals, though such true tradition is only seen in those forms whose high capacity for learning is combined with a higher development of their social life. True tradition has been demonstrated in jackdaws, greylag geese, and rats. But knowledge thus transmitted is limited to very simple things, such as pathfinding, recognition of certain foods and of enemies of the species, and—in rats—knowledge of the danger of poisons. However, no means of communication, no learned rituals are ever handed down by tradition in animals. In other words, animals have no culture.

One indispensable element which simple animal traditions have in common with the highest cultural traditions of man is habit. Indubitably it is habit which, in its tenacious hold on the already acquired, plays a similar part in culture as heredity does in the phylogenetic origin of rites. Once an unforgettable experience brought home to me how similar the basic function of habit can be in such dissimilar processes as the simple formation of path habits in a goose and the cultural development of sacred rites in Man. At the time, I was making observations on a young greylag goose which I had reared from the egg and which had transferred to me, by that remarkable process called imprinting, all the behavior patterns that she would normally have shown to her parents. In her earliest childhood, Martina had acquired a fixed habit: when she was about a week old I decided to let her walk upstairs to my bedroom instead of carrying her up, as until then had been my custom. Greylag geese resent being touched, and it frightens them, so it is better to spare them this indignity if possible. In our house in Altenberg the bottom part of the staircase, viewed from the front door, stands out into the middle of the right-hand side of the hall. It ascends by a right-angled turn to the left, leading up to the gallery on the first floor. Opposite the front door is a very large window. As Martina, following obediently at my heels, walked into the hall, the unaccustomed situation suddenly filled her with terror and she strove, as frightened birds always do, toward the light. She ran from the door straight toward the window, passing me where I now stood on the bottom stair. At the window, she waited a few moments to calm down, then, obedient once more, she came to me on the step and followed me up to my bedroom. This procedure was repeated in the same way the next evening, except that this time her detour to the window was a little shorter and she did not remain there so long. In the following days there were further developments: her pause at the window was discontinued and she no longer gave the impression of being frightened. The detour acquired more and more the character of a habit, and it was funny to see how she ran resolutely to the window and, having arrived there, turned without pausing and ran just as resolutely back to the stairs, which she then mounted. The habitual detour to the window became shorter and shorter, the 180° turn became an acute angle, and after a year there remained of the whole path habit only a right-angled turn where the goose, instead of mounting the bottom stair at its right-hand end, nearest the door, ran along the stair to its left and mounted it at right angles.

One evening I forgot to let Martina in at the right time, and when I finally remembered her it was already dusk. I ran to the front door, and as I opened it she thrust herself hurriedly and anxiously through, ran between my legs into the hall and, contrary to her usual custom, in front of me to the stairs. Then she did something even more unusual: she deviated from her habitual path and chose the shortest way, skipping her usual right-angle turn and mounting the stairs on the right-hand side, "cutting" the turn of the stairs and starting to climb up. Upon this, something shattering happened: arrived at the fifth step, she suddenly stopped, made a long neck, in geese a sign of fear, and spread her wings as for flight. Then she uttered a warning cry and very nearly took off. Now she hesitated a moment, turned around, ran hurriedly down the five steps and set forth resolutely, like someone on a very important mission, on her original path to the window and back. This time she mounted the steps according to her former custom from the left side. On the fifth step she stopped again, looked around, shook herself and greeted, behavior mechanisms regularly seen in greylags when anxious tension has given place to relief. I hardly believed my eyes. To me there is no doubt about the interpretation of this occurrence: the habit had become a custom which the goose could not break without being stricken by fear.

This interpretation will seem odd to some people but I can testify that similar behavior is well known to people familiar with the higher animals. Margaret Altmann, who studied wapiti and moose in their natural surroundings and followed their tracks for months in the company of her old horse and older mule, made very significant observations on her two hoofed collaborators. If she had camped several times in a certain place, she could never afterward move her animals past that place without at least "symbolically" stopping and making a show of unpacking and repacking.

There is an old tragicomic story of a preacher in a small town of the American West, who bought a horse without knowing that it had been ridden for years by a habitual drunkard. The reverend gentleman was forced by his horse to stop at every inn and, in a way analogous to the feigned camping of Margaret Altmann, to go in for at least a few minutes. Thus he fell into disrepute in his parish and finally, in desperation, took to drink himself. This fictitious comedy could, at least with regard to the horse's behavior, be literally true.

To the pedagogue, the psychologist, the ethnologist, and the psychiatrist, the above described behavior pattern of higher animals will seem strangely familiar. Anyone who has children of his own, or has learned how to be a tolerably useful aunt or uncle, knows from experience how tenaciously little children cling to every detail of the accustomed, and how they become quite desperate if a storyteller diverges in the very least from the text of a familiar fairy tale. And anyone capable of self-observation will concede that even in civilized adults habit, once formed, has a greater power than we generally admit. I once suddenly realized that when driving a car in Vienna I regularly used two different routes when approaching and when leaving a certain place in the city, and this was at a time when no one-way streets compelled me to do so. Rebelling against the creature of habit in myself, I tried using my customary return route for the outward journey, and vice versa. The astonishing result of this experiment was an undeniable feeling of anxiety so unpleasant that when I came to return I reverted to the habitual route.

My description will call to the mind of the ethnologist the magic and witchcraft of many primitive peoples; that these are very much alive today even in civilized people can be seen by the fact that most of us still perform undignified little "sorceries" such as "touching wood" or throwing spilled salt over our shoulder.

My examples of animal behavior will remind the psychiatrist and the psychoanalyst of the compulsive repetition of some acts, a symptom of certain types of neurosis. In a mild form, the same phenomenon can be observed in many children. I remember clearly that, as a child, I had persuaded myself that something terrible would happen if I stepped on one of the lines, instead of into the squares of the paving stones in front of the Vienna Town Hall. A. A. Milne gives an excellent impression of this same fancy of a child in his poem "Lines and Squares."

All these phenomena are related. They have a common root in a behavior mechanism whose species-preserving function is obvious: for a living being lacking insight into the relation between causes and effects it must be extremely useful to cling to a behavior pattern which has once or many times proved to achieve its aim, and to have done so without danger. If one does not know which details of the whole performance are essential for its success as well as for its safety, it is best to cling to them all with slavish exactitude. The principle, "You never know what might happen if you don't," is fully expressed in such superstitions.

Even when a human being is aware of the purely fortuitous origin of a certain habit and knows that breaking it does not portend danger, nevertheless an undeniable anxiety impels him to observe it, and gradually the ingrained behavior becomes a custom. So far, the situation is the same in animals as in man. However, a new and significant note is struck from the moment when the human being no longer acquires the habit for himself but learns it from his parents by cultural transmission. First, he no longer knows the reasons for the origin of the particular behavior prescription. The pious Jew or Moslem abhors pork without being conscious that it was insight into the danger of trichinosis which probably caused his lawmakers to impose the prohibition. Second, the revered father-figure of the lawmaker, remote in time as in mythology, undergoes an apotheosis, making all his laws seem godly and their infringement a sin.

The North American Indians have evolved an appeasement ceremony which stirred my imagination in the days when I still played Red Indians: it is the ritual of smoking the pipe of peace, the calumet of friendship. Later, when I knew more about the phylogenetic origin of innate rites, about their aggression-inhibiting action, and above all, about the amazing analogies between the phylogenetic and the cultural origin of symbols, I suddenly visualized the scene that must have taken place when, for the first time, two enemy Indians became friends by smoking a pipe together.

Spotted Wolf and Piebald Eagle, chiefs of neighboring tribes, both old and experienced and rather tired of war, have agreed to make an unusual experiment: they want to settle the question of hunting rights on the island in Little Beaver River, which separates the hunting grounds of their tribes, by peaceful talks instead of by war. This attempt is, the beginning, rather embarrassing, because the wish to negotiate might be misinterpreted as cowardice. Thus when they finally meet, in the absence of their followers, they are both very embarrassed, but as neither dares to admit it, either to himself or to the other, they approach each other in a particularly proud, provocative attitude, staring fixedly at each other and sitting down with the utmost dignity. And then for a long time nothing happens. Anyone who has ever bought a piece of land from an Austrian or a Bavarian farmer knows that whichever one first mentions the matter in hand has already half lost the bargain; and probably the same thing applies to Red Indians. Who knows how long the two chiefs sat face to face?

If you have to sit without moving so much as a face muscle, so as not to betray inner tension, if you are longing to do something but prevented by strong opposing motives from doing it, if in other words you are in a conflict situation, it is often a relief to do a third, neutral thing which has nothing to do with the two conflicting motives and which, moreover, shows apparent indifference to them. The ethologist calls this a displacement activity; colloquially it is called a gesture of embarrassment. All the smokers I know exhibit the same behavior in cases of inward conflict: they put their hand in their pocket, take out their cigarettes or pipe, and light it. Why should the people who invented tobacco-smoking, and from whom we first learned it, do otherwise?

And so Spotted Wolf, or perhaps Piebald Eagle, lighted his pipe, at that time not yet the pipe of peace, and the other chief did the same. Who does not know it, the heavenly, tension-relieving catharsis of smoking? Both chiefs became calmer, more self-assured, and their relaxation led to complete success of the negotiations. Perhaps at the next meeting one of the chiefs lighted his pipe at once, perhaps at the third encounter one had forgotten his pipe and the other—now more tolerant—shared his with him. But perhaps a whole series of countless repetitions of the ceremony was necessary before it gradually became common knowledge that a pipe-smoking Indian is more ready to negotiate than a nonsmoking one. Perhaps it may have taken centuries before the symbol of pipe-smoking unequivocally meant peace. But it is quite certain that in the course of generations the original gesture of embarrassment developed into a fixed ritual which became law for every Indian and prohibited aggression after pipe-smoking. Fundamentally this is the same inviolable inhibition as that which prevented Margaret Altmann's horse from passing the camping site and Martina from missing her customary detour to the window.

However, we would be neglecting an essential side of the matter if we only stressed the inhibiting function of the culturally evolved ritual. Though governed and sanctified by the superindividual, tradition-bound, and cultural superego, the ritual has retained, unaltered, the nature of a habit which is precious to us and to which we cling more fondly than to any habit formed only in the course of an individual life. And herein lies the deep significance of the movement patterns and pageantry of cultural ceremonies. The austere iconoclast regards the pomp of the ritual as an unessential superficiality which even diverts the mind from a deeper absorption in the spirit of the thing symbolized. I believe that he is entirely wrong. If we take pleasure in all the pomp and ceremony of an old custom, such as decorating the Christmas tree and lighting its candles, this presupposes that we love the traditionally transmitted. Our fidelity to the symbol implies fidelity to everything it signifies, and this depends on the warmth of our affection for the old custom. It is this feeling of affection that reveals to us the value of our cultural heritage. The independent existence of any culture, the creation of a superindividual society which outlives the single being, in other words all that represents true humanity is based on this autonomy of the rite making it an independent motive of human action.

The formation of traditional rites must have begun with the first dawning of human culture, just as at a much lower level, phylogenetic rite formation was a prerequisite for the origin of social organization in higher animals. In the following brief description of these two processes I should stress their analogous nature, which is explained by their common functions.

In both cases, a behavior pattern by means of which a species in the one case, a cultured society in the other, deals with certain environmental conditions, acquires an entirely new function, that of communication. The primary function may still be performed, but it often recedes more and more into the background and may disappear completely so that a typical change of function is achieved. Out of communication two new equally important functions may arise, both of which still contain some measure of communicative effects. The first of these is the channeling of aggression into innocuous outlets, the second is the formation of a bond between two or more individuals.

In both cases, the selection pressure of the new function has wrought analogous changes on the form of the primal, non-ritualized behavior. It quite obviously lessens the chance of ambiguity in the communication that a long series of independently variable patterns should be welded into one obligatory sequence. The same aim is served by strict regulation of the speed and amplitude of the motor patterns. Desmond Morris has drawn attention to this phenomenon which he has termed the typical intensity of movements serving as signals. The display of animals during threat and courtship furnishes an abundance of examples, and so does the culturally developed ceremonial of man. The deans of the university walk into the hall with a "measured step"; pitch, rhythm, and loudness of the Catholic priest's chanting during mass are all strictly regulated by liturgic prescription. The unambiguity of the communication is also increased by its frequent repetition. Rhythmical repetition of the same movement is so characteristic of very many rituals, both instinctive and cultural, that it is hardly necessary to describe examples. The communicative effect of the ritualized movements is further increased, in both cases, by exaggerating all those elements which, in the unritualized prototype, produce visual or auditory stimulation while those of its parts that are originally effective in some other, mechanical way are greatly reduced or completely eliminated.

This "mimic exaggeration" results in a ceremony which is, indeed, closely akin to a symbol and which produces that theatrical effect that first struck Sir Julian Huxley as he watched his Great Crested Grebes. A riot of form and color, developed in the service of that particular effect, accompanies both phyletic and cultural rituals. The beautiful forms and colors of a Siamese Fighting Fish's fins, the plumage of a Bird of Paradise, the Peacock's tail, and the amazing colors on both ends of a Mandrill have one and all evolved to enhance some particular ritualized movements. There is hardly a doubt that all human art primarily developed in the service of rituals and that the autonomy of "art for art's sake" was achieved only by another, secondary step of cultural progress.

The direct cause of all these changes which make the instinctive and the cultural ceremonies so similar to each other, indubitably is to be sought in the selection pressure exerted by the limitations of the "receiving set" which must respond correctly and selectively to the signal emanating from the "sender," if the system of communication is to function properly. For obvious reasons, it is the easier to construct a receiver selectively responding to a signal, the more simple and, at the same time, unmistakable the signal is. Of course, sender and receiver also exert a selection pressure on each other's development and may become very highly differentiated in adaptation to each other. Many instinctive rituals, many cultural ceremonies, indeed all the words of all human languages owe their present form to this process of convention between the sender and the receiver in which both are partners in a communicative system developing in time. In such cases, it is often quite impossible to trace back, to an "unritualized model," the origin of a ritual, because its form is changed to a degree that renders it unrecognizable. However, if, in some other living species, or in some still surviving other cultures, some intermediate steps on the same line of development are accessible to study, comparative investigation may still succeed in tracing back the path along which the present form of some bizarre and complicated ceremony has come into being. This, indeed, is one of the tasks that make comparative studies so fascinating.

Both in phylogenetic and in cultural ritualization the newly evolved behavior patterns achieve a very peculiar kind of autonomy. Both instinctive and cultural rituals become independent motivations of behavior by creating new ends or goals toward which the organism strives for their own sake. It is in their character of independent motivating factors that rituals transcend their original function of communication and become able to perform their equally important secondary tasks of controlling aggression and of forming a bond between certain individuals. On page 64 we have already learned in what way a ceremony becomes a bond; in Chapter Ten, I shall explain in some detail how an aggression-controlling ceremony can develop into a strong bond comparable to human love and friendship.

In cultural ritualization, the two steps of development leading from communication to the control of aggression and, from this, to the formation of a bond, are strikingly analogous to those that take place in the evolution of instinctive rituals, as illustrated in Chapter Ten by the triumph ceremony of geese. The triple function of suppressing fighting within the group, of holding the group together, and of setting it off, as an independent entity, against other, similar units, is performed by culturally developed ritual in so strictly analogous a manner as to merit deep consideration.

Any human group which exceeds in size that which can be held together by personal love and friendship, depends for its existence on these three functions of culturally ritualized behavior patterns. Human social behavior is permeated by cultural ritualization to a degree which we do not realize for the very reason of its omnipresence. Indeed, in order to give examples of human behavior which, with certainty, can be described as nonritualized, we have to resort to patterns which are not supposed to be performed in public at all, like uninhibited yawning and stretching, picking one's nose or scratching in unmentionable places. Everything that is called manners is, of course, strictly determined by cultural ritualization.

"Good" manners are by definition those characteristic of one's own group, and we conform to their requirements constantly; they have become "second nature" to us. We do not, as a rule, realize either their function of inhibiting aggression or that of forming a bond. Yet it is they that effect what sociologists call "group cohesion."

The function of manners in permanently producing an effect of mutual conciliation between the members of a group can easily be demonstrated by observing what happens in their absence. I do not mean the effect produced by an active, gross breach of manners, but by the mere absence of all the little polite looks and gestures by which one person, for example on entering a room, takes cognizance of another's presence. If a person considers him- or herself offended by members of his group and enters the room occupied by them without these little rituals, just as if they were not there, this behavior elicits anger and hostility just as overt aggressive behavior does; indeed, such intentional suppression of the normal appeasing rituals is equivalent to overt aggressive behavior.

Aggression elicited by any deviation from a group's characteristic manners and mannerisms forces all its members into a strictly uniform observance of these norms of social behavior. The nonconformist is discriminated against as an "outsider" and, in primitive groups, for which school classes or small military units serve as good examples, he is mobbed in the most cruel manner. Any university teacher who has children and has held positions in different parts of a country, has had occasion to observe the amazing speed with which a child acquires the local dialect spoken in the region where it has to go to school. It has to, in order not to be mobbed by its schoolfellows, while at home it retains the dialect of the family group. Characteristically, it is very difficult to prevail on such a child to speak, in the family circle, the "foreign language" learned at school, for instance in reciting a poem. I believe that the clandestine membership of another than the family group is felt to be treacherous by young children.

Culturally developed social norms and rites are characteristics of smaller and larger human groups much in the same manner as inherited properties evolved in phylogeny are characteristics of subspecies, species, genera, and greater taxonomic units. Their history can be reconstructed by much the same methods of comparative study. Their divergence in historical development erects barriers between cultural units in a similar way as divergent evolution does between species; Erik Erikson has therefore aptly called this process pseudo-speciation.

Though immeasurably faster than phylogenetic speciation, cultural pseudo-speciation does need time. Its slight beginnings, the development of mannerisms in a group and discrimination against outsiders not initiated to them, may be seen in any group of children, but to give stability and the character of inviolability to the social norms and rites of a group, its continued existence over the period of at least a few generations seems to be necessary. For this reason, the smallest cultural pseudo-subspecies I can think of is the school, and it is surprising how old schools preserve their pseudo-subspecific characters throughout the years. The "old school tie," though often an object of ridicule nowadays, is something very real. When I meet a man who speaks in the rather snobbish nasal accent of the old Schotten-Gymnasium in Vienna, I cannot help being rather attracted to him; also I am curiously inclined to trust him just as I myself would probably be more meticulously fair in my social behavior to a man of my old school group than to an outsider.

The important function of polite manners can be studied to great advantage in the social interaction between different cultures and subcultures. A considerable proportion of the mannerisms enjoined by good manners are culturally ritualized exaggerations of submissive gestures most of which probably have their roots in phylogenetically ritualized motor patterns conveying the same meaning. Local traditions of good manners, in different subcultures, demand that a quantitatively different emphasis be put on these expression movements. A good example is furnished by the attitude of polite listening which consists in stretching the neck forward and simultaneously tilting the head sideways, thus emphatically "lending an ear" to the person who is speaking. The motor pattern conveys readiness to listen attentively and even to obey. In the polite manners of some Asiatic cultures it has obviously undergone strong mimic exaggeration; in Austrians, particularly in well-bred ladies, it is one of the commonest gestures of politeness; in other Central European countries it appears to be less emphasized. In some parts of northern Germany it is reduced to a minimum, if not absent. In these subcultures it is considered correct and polite for the listener to hold the head high and look the speaker straight in the face, exactly as a soldier is supposed to do when listening to orders. When I came from Vienna to Königsberg, two cities in which the difference of the motor pattern under discussion was particularly great, it took me some little time to get used to the polite listening gesture of East Prussian ladies. Expecting a tilt of the chin, however small, from a lady to whom I was speaking, I could not help feeling that I had said something shocking when she sat rigidly upright looking me in the face.

Of course the meaning of any conciliatory gesture of this kind is determined exclusively by the convention agreed upon by the sender and the receiver of one system of communication. Between cultures in which this convention is different, misunderstandings are unavoidable. By East Prussian standards a polite Japanese performing the "ear-tending" movement would be considered to be cringing in abject slavish fear, while by Japanese standards an East Prussian listening politely would evoke the impression of uncompromising hostility.

Even very slight differences in conventions of this kind may create misinterpretation of culturally ritualized expression movements. Latin peoples are very often considered as "unreliable" by Anglo-Saxons and Germans, simply because, on a basis of their own convention, they expect more social good will than actually lies behind the more pronounced "effusive" motor patterns of conciliation and friendliness of the French or the Italians. The general unpopularity of North Germans and particularly Prussians in Latin countries is, at least partly, due to this type of misunderstanding. In polite American society I have often suspected that I must give the impression of being rather rude, because I find it difficult to smile quite as much as is demanded by American good manners.

Indubitably, little misunderstandings of this kind contribute considerably to inter-group hate. The man who, in the manner described, has misinterpreted the social signals of a member of another pseudo-subspecies, feels that he has been intentionally cheated or wronged. Even the mere inability to understand the expression movements and rituals of a strange culture creates distrust and fear in a manner very easily leading to overt aggression.

From the little peculiarities of speech and manner which cause the smallest possible subcultural groups to stick together, an uninterrupted gradation leads up to the most elaborated, consciously performed, and consciously symbolical social norms and rites which unite the largest social units of humanity in one nation, one culture, one religion, or one political ideology. Studying this system by the comparative method, in other words, investigating the laws of pseudo-speciation, would be perfectly possible, though more complicated than the study of speciation, because of the frequent overlapping of group concepts, as for instance of the national and the religious units.

I have already said that an emotional appreciation of values gives motivational power to every ritualized norm of social behavior. Erik Erikson has recently shown that the conditioning to the distinction of good and bad begins in early babyhood and continues all through the ontogeny of a human being. In principle, there is no difference between the rigidity with which we adhere to our early toilet training and our fidelity to the national or political norms and rites to which we become object-fixated in later life. The rigidity of the transmitted rite and the tenacity with which we cling to it are essential to its indispensable function. At the same time, like the corresponding function of even more rigid instinctive patterns of social behavior, they need supervision by our rational, responsible morality.

It is perfectly right and legitimate that we should consider as "good" the manners which our parents have taught us, that we should hold sacred the social norms and rites handed down to us by the tradition of our culture. What we must guard against, with all the power of rational responsibility, is our natural inclination to regard the social rites and norms of other cultures as inferior. The dark side of pseudo-speciation is that it makes us consider the members of pseudo-species other than our own as not human, as many primitive tribes are demonstrably doing, in whose language the word for their own particular tribe is synonymous with "Man." From their viewpoint it is not, strictly speaking, cannibalism if they eat the fallen warriors of an enemy tribe. The moral of the natural history of pseudo-speciation is that we must learn to tolerate other cultures, to shed entirely our own cultural and national arrogance, and to realize that the social norms and rites of other cultures, to which their members keep faith as we do to our own, have the same right to be respected and to be regarded as sacred. Without the tolerance born of this realization, it is all too easy for one man to see the personification of all evil in the god of his neighbor, and the very inviolability of rites and social norms which constitutes their most important property can lead to the most terrible of all wars, to religious war—which is exactly what is threatening us today.

Here, as so often when discussing human behavior from the viewpoint of natural science, I am in danger of being misunderstood. I did indeed say that man's fidelity to all his traditional customs is caused by creature habit and by animal fear at their infraction. I did indeed emphasize the fact that all human rituals have originated in a natural way, largely analogous to the evolution of social instincts in animals and man. I have even stressed the other fact that everything which man by tradition venerates and reveres, does not represent an absolute ethical value, but is sacred only within the frame of reference of one particular culture. However, all this does not in any sense derogate from the unfaltering tenacity with which a good man clings to the handed-down customs of his culture. His fidelity might seem to be worthy of a better cause, but there *are* few better causes! If social norms and customs did not develop their peculiar autonomous life and power, if they were not raised to sacred ends in themselves, there would be no trustworthy communication, no faith, and no law. Oaths cannot bind, nor agreements count, if the partners to them do not have in common a basis of ritualized behavior standards at whose infraction they are overcome by the same magic fear as seized my little greylag on the staircase in Altenberg.

Chapter Six

The Great Parliament of Instincts

As we have learned in the previous chapter, the phylogenetic process of ritualization creates a new autonomous instinct which interferes as an independent force in the great constitution of all other instinctive motivations. Its primary function, which consists, as we know, of a communication, can prevent the harmful effects of aggression by inducing mutual understanding between members of a species. It is not only in man that a quarrel often arises because one person mistakenly thinks that the other means harm to him; in this connection the rite is tremendously important to our theme. In addition, as we have seen in the example of the triumph ceremony in geese, ritual can achieve such power as an independent motivation that it can successfully oppose the might of aggression in the great parliament of instincts. In order to explain how ritual checks the aggressive drive without really weakening it or hindering its species-preserving function, I must say something about the constitution of instincts. This constitution resembles a parliament, since it is a more or less complete system of interactions between many independent variables; its true democratic nature has developed through a probationary period in evolution, and it produces, if not always complete harmony, at least tolerable and practicable compromises between different interests.

What is an instinct? The terms often used for various instinctual motivations are frequently tainted by the unfortunate heritage of "finalistic" thinking. A "finalist," in this bad sense of the word, is someone who confuses the question "What for?" with the question "How come?" and thus believes that by demonstrating the species-preserving reason for a certain function he has solved the problem of its causation. As the determinant of a concrete function whose survival value is obvious, such as eating, copulation, or flight, it is tempting to postulate a special impulse or instinct. We are all familiar with the term "reproductive instinct." However, we should not imagine—as many vitalistic students of instinct did—that the invention of such a term provides the explanation of the process in question. The conceptions corresponding to such labels are no better than those of nature's abhorrence of the vacuum or the "phlogiston" which are only names for a process but "fraudulently pretend to contain an explanation of it," as John Dewey has bluntly put it. Since we are attempting in this book to find the causal explanation for the functional failure of a certain instinct, aggression, we must not limit ourselves to the investigation of its "What for?" as we did in the third chapter. We must gain insight into its normal causes so as to understand the nature of its disorders and perhaps learn to repair them.

A definite and self-contained function of an organism, such as feeding, reproduction, or self-preservation, is never the result of a single cause or of a single drive. The explanatory value of a concept such as "reproductive instinct" or "instinct of self-preservation" is as null as the concept of an "automobile force," which I could use just as legitimately to explain the fact that my ancient car still goes. But anyone who knows —and pays for—the repairs which keep it going will not be tempted to believe in such mystic power—and it is the repairs with which we are here concerned! The neurophysiological organization which we call instinct functions in a blindly mechanical way, particularly apparent when its function goes wrong. Such functional failures must be experienced in order to appreciate the folly of believing in instinct with a capital "I," and regarding it as a preternatural, direction-giving "factor" neither in need of nor accessible to a causal explanation.

A functionally uniform behavior pattern such as feeding or reproduction is always achieved by a very complicated interaction of many physiological causes, whose systemic function has been "invented" and thoroughly tested by the two constructors of evolution, mutation and selection. The several physiological causes entering into this interaction are sometimes related to each other in a balanced reciprocal influence: sometimes one influences the others more than it is influenced in return, and some are relatively independent of the whole working structure and influence it more than they are influenced by it. Good examples of such "relatively independent building stones" are the elements of the skeleton.

In the realm of behavior, the hereditary co-ordinations or instinct movements are independent building stones. As unchangeable in their form as the hardest skeletal component, each one cracks its own whip over the organism as a whole. As we already know, each one speaks up when it has been silent for too long, and forces the animal or human to get up and search actively for the special set of stimuli which elicit it and no other hereditary co-ordination. It is therefore a mistake to suppose that an instinctive movement whose species-preserving function serves, for example, nutrition, must necessarily be caused by hunger. We know that our own dogs go through the motions of smelling, seeking, chasing, biting, and shaking to death with equal enthusiasm whether they are hungry or not, and every dog owner knows that a dog which is a passionate hunter cannot be cured of its passion by abundant feeding. The same applies to the instinctive movements of preying in the cat, to the well-known prying movement of the starling, which is almost continuously performed independently of the bird's state of hunger, in short to all the little servants of species preservation such as running, flying, gnawing, pecking, grooming, etc. Each of these hereditary co-ordinations has its own spontaneity and causes its own appetitive behavior.

Are these little partial drives completely independent of each other? Do they form a mosaic which owes its functional completeness only to the construction of evolution? In extreme cases this may be so, and not very long ago these special cases were considered to be the rule. In the early days of comparative behavior research it was thought that one drive at a time exclusively governed the whole animal. At that time Julian Huxley used a good metaphor, which I myself quoted for years in my lectures. He compared the human being or animal to a ship commanded by many captains. In the human, all these commanders are on the bridge at the same time and each voices his opinion. In doing so they sometimes reach a wise compromise which provides a better solution to their problems than the single opinion of the cleverest among them; but sometimes they cannot agree and then the ship is without any rational leadership. In the animal, on the other hand, the captains keep to an agreement that only one at a time will stand on the bridge. This last simile is very apt in some cases of animal behavior in conflict situations, and it is understandable that at one time we overlooked the fact that these are only relatively rare, special cases. Moreover, it is the simplest form of interaction between two conflicting drives when one simply suppresses or supplants the other. In those early days it was entirely legitimate and right to concentrate on the simplest and most easily analyzed processes, even if they were not the most common.

In reality, all imaginable interactions can take place between two impulses which are variable independently of each other. One impulse can one-sidedly aid and accelerate the other; the two can support each other; the motor effects of two independent sources of impulses can, without otherwise bearing any relation to each other, be superimposed on each other in one and the same behavior pattern. Finally, in addition to many further forms of interaction, the enumeration of which would carry us too far, there is the one rare, special case which really corresponds to Huxley's allegory: of two impulses, either can eliminate the other in a sudden and complete switch-off effect, dependent only on which of the two drives happens to be momentarily stronger. There is only one drive of which it can be said that it generally subjugates all others—the escape drive—but even this one sometimes meets its master.

The everyday, common, "cheap," fixed motor patterns which I have called the "little servants of species preservation" are often at the disposal of more than one of the "big" drives. Particularly the behavior patterns of locomotion, such as running, flying, swimming, etc., also those of pecking, gnawing, and digging, can be used in the service of feeding, reproduction, flight, and aggression, which we will here call the "big drives." Because the little servants play a subsidiary part of "common final pathways" to various superior systems, in particular to the above-mentioned "big four," I have called them tool activities. This, however, does not mean that such motor patterns lack their own spontaneity. On the contrary, it is compatible with a widespread principle of natural economy that, for example in a dog or a wolf, the spontaneous production of the separate impulses of sniffing, tracking, running, chasing, and shaking to death is roughly adapted to the demands of hunger. If we exclude hunger as a motive, by the simple method of keeping the dish full, it will soon be noticed that the animal sniffs, tracks, runs, and chases hardly less than when these activities are necessary to allay its hunger. Still, if the dog is very hungry, he does all this quantitatively more. Thus, though the tool instincts possess their own spontaneity, they are driven, in this case by hunger, to perform more than they would if left alone. Indeed, a drive can be driven.

This driving of an inherently spontaneous function by a stimulus from elsewhere is not new or even rare in physiology. An instinctive movement pattern is a reaction in so far as it can be elicited by the impulse of an exogenous stimulus or by another, independent endogenous drive. Only in the absence of such stimuli does it show its own spontaneity.

An analogous process is well known in the stimulus-producing centers of the heart; normally the heartbeat is elicited by the rhythmic-automatic stimuli of the sinus node, an organ of highly specialized muscle tissue situated near the entrance of the blood stream in the atrium of the heart. Further down, in the direction of the blood stream, at the junction with the ventricle, there is a second such organ, the atrioventricular node, to which a bundle of stimulus-carrying muscle fibers leads from the sinus nodes. Both nodes produce stimuli that cause the heart chambers to contract. The sinus node works quicker than the atrioventricular node, thus the latter is never in a position to "behave" spontaneously, since every time it gets ready to discharge a stimulus, it receives an impulse from its superior causing it to discharge just a bit sooner than it would have done by itself. So the superior imposes on its subordinate its own working rhythm. If we make the classical experiment of Stannius and interrupt the connection between the nodes by ligaturing the stimulus-conducting bundle, the atrioventricular node is freed from the tyranny of the sinus node and does what the subordinate very often does in such cases: it stops working, in other words the heart stops for a moment. This is called the preautomatic pause. After a short rest, the atrioventricular node "notices" that it can itself produce stimuli and discharge them. Hitherto it had not been able to do so because fractions of seconds beforehand it had always received an impulse from above.

A relationship exactly analogous to that of the atrioventricular node to the sinus node exists in most fixed motor patterns to various superior sources of motivation. Conditions are here complicated by the fact that, first, one servant often has several masters, and secondly, these masters may be of an extremely different nature. They may, like the sinus node, be automatic-rhythmic, stimulus-producing organs; they may be internal and external receptors receiving endogenous and exogenous stimuli, including tissue needs such as hunger, thirst, or lack of oxygen, which they relay toward the center in the form of excitation. Finally they may be endocrine glands whose hormones stimulate definite nervous processes ("hormone" comes from the Greek *hormao,* I drive); but in every case, the activity ruled from a higher position does not bear the character of a "reflex," that is to say, the whole organization of instinctive motor co-ordinations does not behave like a machine which, as long as it is not used, remains passive for an unlimited time and "waits" till somebody presses the button of its elicitation. It resembles a horse which may need bit and spurs to make it obey its master, but which must have daily exercise to keep down its superfluous energy. This energy, in the case of the instinct with which we are here concerned—intra-specific aggression—may become very dangerous.

As already mentioned, the amount of spontaneous production of a particular instinctive movement is always more or less adapted to the expected requirements. Sometimes it is expedient if this stimulus production is doled out economically, for example in the stimulus production of the atrioventricular node, for if this produces more than the sinus node "orders," the result is extra systole, a condition disagreeably familiar to nervous people and consisting in the interruption of the normal heartbeat by an arhythmic contraction of the ventricle. In other cases, a constant overproduction may be harmless or even useful. When a dog runs more than is necessary to catch a prey, or a horse bucks and kicks for no apparent reason, these motor patterns of flight and self-defense are good practice for the event of real danger.

The available overproduction of tool activities will be greatest where it is least predictable how much of the activity will be needed in a particular case before the species-preserving function is accomplished. A preying cat may at one time be forced to lie in wait for several hours in front of a mousehole, and at another to catch a mouse which has fortuitously crossed her path, by a quick pounce without any lurking or stalking. In general, however, as may be observed out-of-doors in the country, a cat has to stalk and lie in wait patiently for a long time before it is at last able to consummate the end actions of killing and eating its prey. In watching such chains of action, it is all too easy to make a wrong comparison with human purposive behavior. Involuntarily one tends to assume that the cat performs the movement pattern of prey catching "for the sake of eating only." That this is not so can be demonstrated experimentally. Leyhausen gave to cats which were keen hunters one mouse after the other and observed the order in which the part actions of preying and eating disappeared. First the cat stopped eating but killed a few more mice, leaving them untouched. Next the killing bite disappeared, but the cat continued to stalk and to catch the mice. Later still, when the movement pattern of catching was exhausted, the cat still did not stop stalking the mice and indeed, in so doing, it always chose those farthest away in the opposite corner of the room, and ignored those that ran over its forepaws.

In this experiment it can be calculated how often every single one of the described part actions was performed before it was exhausted, and the resulting figures bear an obvious relation to those of average, everyday use. Obviously a cat must very often stalk its prey before it comes so near that an attempt to catch it has any promise of success, and only after many catching attempts can it seize the prey in its claws and inflict the lethal bite. This does not always succeed the first time, so several killing bites must always be in reserve for each eating of a mouse.

Whether one of the part actions is performed under its own impulse alone, or additionally under that of another one, and which one that is, depends in complex behavior mechanisms like this on external conditions determining the "demand" for every single behavior pattern. To the best of my knowledge this was first clearly expressed by the child psychiatrist and psychologist, René Spitz. He observed in human sucklings that, if their milk was too easily sucked from the bottle, even after they were satisfied and had rejected the nipple they still had an excess of sucking movements which they used up on substitute objects. Geese show similar behavior in the activities of eating and food seeking, if they are kept in a pond where there is no food obtainable by the movement pattern of so-called "up-ending." If the birds are fed only on the shore, sooner or later it will be observed that they perform up-ending movements for their own sake. If they are now fed to the point of satiety, still on the shore, with a certain type of corn and this corn is then thrown into the water, the birds begin to up-end and they really eat what they fetch up. It may be said that "they eat to up-end." We can also make the opposite experiment and let the geese acquire their whole nourishment by strenuous up-ending in deeper water. If they are allowed to eat in this way until they stop, and then are given the same food on dry land, they eat a considerable amount, thus demonstrating that in this case they only "up-ended to eat." Thus it is generally not possible to state which of two spontaneous, motivation-producing organizations "drives" or "dominates" the other one.

So far we have only spoken of the interaction between those partial drives which co-operate in a common function, in our examples, feeding. Rather different is the relation between impulses which each have a different function and thus belong to different instinct organizations. In this case, it is not mutual driving or support that is the rule, but a relationship of rivalry, each of the impulses trying to "assert its right." As Erich von Holst has shown, even on the plane of the smallest muscle contractions several stimulus-producing elements not only vie with each other but also form sensible compromises by methodical mutual influence. This influence consists roughly in the fact that each of two endogenous rhythms strives to force upon the other its own frequency and to keep it in a constant phase-relation. The fact that all nerve cells whose impulses cause contraction of one muscle generally discharge simultaneously is the consequence of this mutual influence.

On the somewhat higher integration level of the movement of an extremity, for example of a fish's fin, the same processes effect an expedient alternating play between antagonistic muscles, that is those that move the particular limb alternately in opposite directions of space. Every rhythmical to-and-fro movement of a fin, a leg, or a wing, such as we see everywhere in animal locomotion, is caused by the alternating prevalence of opposed impulses, and this applies both to the muscles involved and to the stimulus-producing centers of the nervous system. The movement is always the result of a "conflict" of independent and rival impulses whose energies are guided by the laws of "relative co-ordination," as von Holst called these processes of mutual influence, into ordered paths serving the good of the whole organism.

I have no great sympathy for the Greek sage who asserts that war is the father of all things, but with a better right this honorific might be given to conflict. Conflict between independent sources of impulse is able to produce, within the organism, tensions which lend firmness to the whole system, much as the stays of a mast give it stability by pulling in opposite directions. This applies not only to simple performances like the fin stroke of fishes, in which Erich von Holst discovered the laws of relative co-ordination, but also to very many other impulses which are forced by these well-tried parliamentary rules to unite their individual votes in a harmony serving the whole organism.

A simple example of such a conflict is represented by the way in which a dog moves its facial muscles when torn between the drives of fight and flight. The resulting expression, which is generally called threatening, occurs only when the tendency to attack is inhibited by fear, even a very small portion of fear. Without this, the animal bites without threatening, with the calm face which is portrayed in the left upper corner of Figure 3 and which betrays only slight tension of a kind similar to that shown when the dog is eying the food carried by his owner. If the reader understands dogs, he should try to interpret for himself the forms of expression shown in the figure. He should try to envisage the situations in which his dog makes each particular face. Then, as a second exercise, he should try to predict what the animal will do in the next second or minute.

I will now give the solution to the problem: in the case of the dog in the center of the top row, I should say he is facing an equally strong, respected, but scarcely feared rival who dares to take action as little as he does himself, and my behavior prediction would be that both will keep this position for minutes on end, then slowly, "saving face," move away from each other, and finally, at some distance from each other, simultaneously lift their legs. The dog at the top right fears his rival even less; the encounter may proceed as above but it may

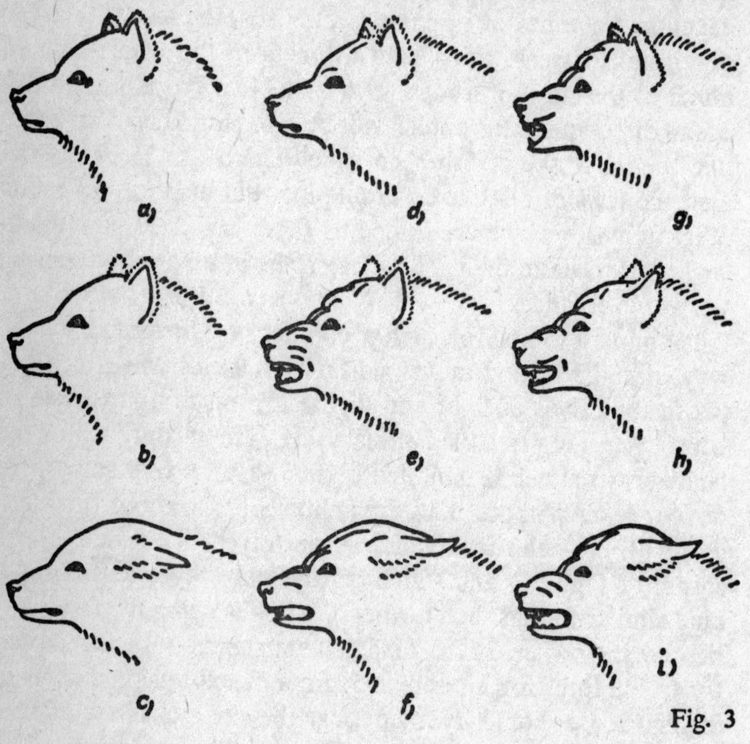

Fig. 3

also—particularly if one dog is unsure of itself—break out into a nasty, noisy fight. Every reader who is intelligent enough to have read thus far will already have noticed that the dog portraits in the diagrams follow in a certain order: aggression increases toward the right and fear increases vertically.

The explanation and prediction of behavior is easiest in the most extreme cases, particularly in the facial expression portrayed in the bottom right-hand corner: *such* rage combined with *such* fear can be seen only in one situation, namely if the dog is facing a hated or greatly feared enemy at a short distance and is unable, for some reason or other, to flee. I can only imagine two situations in which this can occur: either the dog is fixed mechanically to the spot, perhaps cornered or trapped, or it is a bitch defending her young from an approaching enemy. There is also the romantic possibility that a particularly faithful dog is defending its sick or wounded master. In any case it is quite clear what is going to happen: if the enemy, however powerful he may be, advances one step, there follows the desperate attack of which we have already spoken—the critical reaction.

My dog-loving reader has just done what the ethologists N. Tinbergen and Jan van Iersel call making a motivation analysis. This consists basically of three steps in which we draw our information from three sources. First, we try to test the situation for its content of stimuli of different significance. Is my dog afraid of the other, and if so how much? Does he hate him, or does he respect him as an older friend and "pack leader"? and other similar questions. Secondly, we try to split the observed movement into its component parts. We see in our diagram how the flight impulse draws the ears and the corners of the mouth backward and downward, while aggression leads to lifting of the upper lip and opening the mouth, both of which actions are preparations, "intention movements," to bite. These movements or attitudes can be analyzed quantitatively: one could measure their extent and literally state that this or that dog shows so and so many parts of an inch of fear or of anger. Thirdly, we can evaluate the behavior patterns which follow the motor patterns thus analyzed. If the opinion we have formed from situation analysis and movement analysis is correct and the right-hand upper dog is only angry and hardly afraid, attack but almost never flight will follow the expression movement depicted. If it is correct that, in the center dog, anger and fear are almost equally mixed, this expression will be followed by attack in half the cases and by flight in the other half. Tinbergen and his collaborators made numerous motivation analyses on suitable objects, particularly on the threatening movements of sea-gulls, and the conformity of agreement from these three sources has convincingly proved, on the broadest statistical basis, the correctness of their conclusions.

When young students, familiar with animals, are introduced to the technique of motivation analysis, they are often disappointed that their painstaking analysis, above all the dull statistical evaluation, finally shows nothing beyond what a sensible person with eyes in his head and a good knowledge of animals knows already. There is, however, a difference between seeing and proving, and it is this difference which divides art from science. To the artistic observer, the scientist who seeks proof seems a pitiable wretch, and conversely, the use of mere perception as the source of knowledge seems highly suspicious to some scientists. There is, in fact, a school of orthodox American behaviorists who seriously attempt to exclude direct observation of animals from their methods. It is a worthwhile task to prove what we have seen, in such a way that these and other "eyeless" but intelligent people are bound to believe it.

Furthermore, statistical analysis may call our attention to irregularities which perception has overlooked. Perception has the function of discovering laws, and it always sees things as rather more beautiful and according to rule than they really are. Hence the solution which it suggests often bears the character of a very "elegant" but rather too simplified hypothesis. The rational analysis of motivation quite often succeeds in showing up the deficiencies of perception.

The greater part of all motivation analyses have so far been concerned with behavior patterns in whose origin only two conflicting drives participate, and these are usually two of the "big four"—hunger, love, fight, and flight. At the present modest stage of our knowledge it is quite legitimate to choose the simplest cases possible for the study of drive conflicts, just as the classicists of behavior research were justified in keeping to those cases in which the animal was influenced by a single drive. However, it is important to realize that behavior determined by only two drive components is almost as rare as that caused by the impulse of a single instinct, acting alone and uninfluenced.

In looking for a favorable object for an exact motivation analysis, a form of behavior should be chosen in which only two equivalent instincts participate. To achieve this end, one can sometimes employ a technical trick such as my collaborator, Helga Fischer, used while making a motivation analysis of the threatening of Greylag geese. In the home surroundings of our Greylags, it proved impossible to reproduce the combination of aggression and flight in their purest state since, in the expression movements of these birds, too many other motivations, particularly sexual ones, "spoke up." On the other hand, a few chance observations showed that the voice of sexuality was almost entirely silenced when the geese were in strange surroundings. There they behaved rather like a migratory flock, kept much closer together, and were more easily frightened; in their social disagreements the effects of the two instincts under examination could be observed in a much purer form. Using food as bait, Helga Fischer trained her geese, all of which were distinguished individually by colored rings, to fly out at her orders to unfamiliar localities well outside the boundaries of our institute and seek their food there. Then she carefully recorded the encounters between certain randomly chosen individuals, mostly ganders, with certain other members of the flock. Since in years of observation she had become thoroughly acquainted with every minute detail in the social rank of her flock, she had an excellent opportunity of making an exact situation analysis of each of these quarrels, so much so that she could often predict what would happen from the

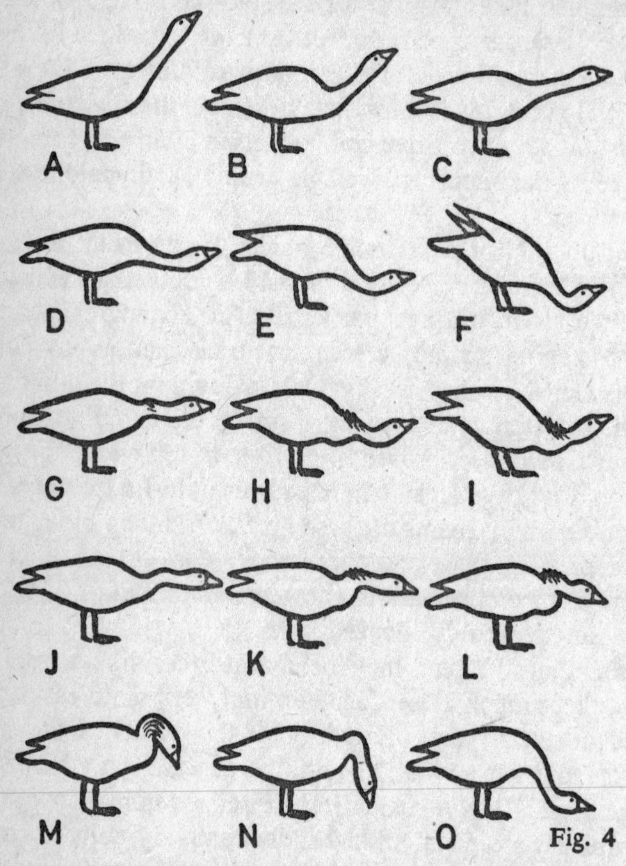

Fig. 4

relative social status of any two opponents. Subsequently, the analysis of the composite movements was made as follows. While making her observations, Helga Fischer used the diagram shown in Figure 4, drawn by our artist Hermann Kacher from photographs of antagonistic encounters of which also exact records had been made. The diagram comprises, in a sufficient number of gradations, all forms of expression movements arising from the conflict between the drives of escape and of aggression. In using this "sampler," the observer need, for example, only note, "Gander Max makes G at Moritz who responds with an E." It was seldom necessary to describe a movement or an attitude as an intermediate between two of the figures.

Situation analysis, movement analysis, and the observation of subsequent behavior were in perfect agreement in showing that the movements depicted in the lines D–F, G–I, and J–L were indeed the expressions of aggression and escape drive. In both sequences, the pictures at the left represent pure aggression, while escape motivation increases from left to right. The difference between the three sequences is explicable by the highest intensity of both drives in the attitude depicted in the upper sequence, and the lowest in the bottom line.

Even in a situation calculated to produce in as pure a form as possible the two motivations of intra-specific aggression and escape, some attitudes are observed which cannot be explained as simple superpositions and/or mixtures of the motor impulses of aggression (pushing the neck forward as in D) and of escape drive (pushing it far back as in M). A, B, and C, as well as N and M, clearly contain some other elementary components. In both cases, other independently variable motivations are involved, the discussion of which would lead us too far. However, it is important for the principle of motivation analysis that there is no other way of ascertaining the number of independent variables involved than by first trying a tentative explanation implying as few independent variables as possible and subsequently introducing additional ones, as the necessity arises. Before proceeding further in an attempt at motivation analysis, the first basic question to be answered is that of the number and the properties of all the independently variable drives which take part in motivating composite behavior. To solve this problem, several investigators, among them P. Wiepkema, have successfully used the exact method of factor analysis.

A good example of motivation analysis in which there are three chief components to be considered was provided by my pupil B. Oehlert in her thesis. The subject of this investigation was the behavior shown by certain cichlids when two previously unacquainted individuals were brought together. Species were chosen in which there is scarcely any external difference between the male and female; in such species any two unacquainted individuals always react to each other with behavior patterns motivated simultaneously by the drives of flight, aggression, and sexuality. In these fish, the behavior patterns arising from every single one of these drive sources can be clearly differentiated, since they manifest themselves, even at their lowest intensity, by their different spatial directions: all sexually motivated movement patterns, including the digging of the nest hollow, cleaning of the spawning stone, and the movements of spawning and fertilizing, are directed toward the ground; all movements of flight, including its slightest intimation, point away from the opponent and usually upward toward the surface; while all movements of aggression, except certain threatening movements containing a component motivated by escape, point toward the opponent. If we know these basic facts and moreover the special motivations of several ritualized expression movements, we are in a particularly favorable position to ascertain quantitively to what extent and in what proportion the separate impulses take part in determining the fish's behavior at a given moment. This analysis is further aided by the fact that many of these fish take on different, easily recognizable color patterns, each of which is characteristic of one of the three important motivations: aggression, sexuality, and escape.

As an unexpected side result of this motivation analysis, B. Oehlert discovered a mechanism of "sex recognition" evidently to be found not only in these fish but in a great many other vertebrates too. In the cichlids under examination male and female are exactly alike, not only externally but also in their movement patterns, even those of the sexual act, fertilization, and oviposition, and it was therefore difficult to find out what mechanisms were at work to prevent homosexual pair formation. One of the greatest demands put to the observational powers of a behavior investigator is that he or she must notice when otherwise widespread behavior patterns do not occur in a certain animal or group. For example, that birds and reptiles lack the motor pattern of yawning—opening wide the mouth and inhaling deeply—is a taxonomically important observation which no zoologist before Heinroth had ever made.

The discovery that the lack of a certain behavior pattern in the male and of another in the female was responsible for the sex recognition in Cichlids entailed particularly acute observation. In these fish, the relation of misceability of the three great drive sources is different in the male and in the female: in the male, the motivations of flight and of sexuality cannot be mixed. If the male has even the slightest fear of his partner, his sexuality is completely extinguished. In the female, there is the same relation between aggression and sexuality: if she is so little in awe of her partner that her aggression is not entirely suppressed, she does not react to him sexually at all. She becomes a Brunhilde and attacks him the more ferociously the more potentially ready she is for sexual reactions, that is, the nearer she is to spawning, in respect of her ovarian and hormonal state.

Conversely, aggression and sexuality are quite compatible in the male; he can treat his partner roughly, chase her all around the tank, and betweenwhiles perform sexual movements and all possible mixed forms of motor patterns. The female may fear the male considerably without suppression of sexually motivated behavior pattern. The bride-to-be may flee before the male and at the same time make use of every breathing space to perform sexually motivated courtship movements. These mixed forms of behavior patterns of flight and sexuality have become, by ritualization, widespread ceremonies which are often called "coyness behavior" and which possess a very definite expression value.

Since this relation of the misceability of the three great drives is different in the two sexes, a male can only pair with an awe-inspired and therefore submissive female, and a female only with an awe-inspiring and therefore dominant male. Thus the behavior mechanism just described guarantees the pairing of two individuals of opposite sex. In many variations, and modified by different ritualizations, this process of sex recognition and pair formation plays an important part in very many vertebrates right up to man. It provides an impressive example of the indispensable species-preserving function that aggression may fulfill in the harmonious play of interactions with other motivations. In addition, it shows how different the relations between the "big" drives can be even in males and females of the same species. Two motives, which in one sex scarcely inhibit each other, exclude each other in the other sex in a sharp shunting mechanism.

As already mentioned, it is an error to assume that the "big four"—hunger, sexuality, flight, and aggression—are irresistible tyrants whose commands brook no contradiction. It is not even true that the more widely distributed and phylogenetically older sources of motivation always dominate over more specialized and more recently evolved instincts. Some of those special drives which guarantee a permanent aggregation of social animals rule the individual so strongly that under certain conditions they can supersede all other drives. The sheep that leaps over the precipice after the leader ram has become proverbial. A greylag goose that has become separated from the flock does everything in its power to find it again, and the drive toward the flock can even overcome the escape drive. Wild geese have repeatedly joined our tame ones in the immediate neighborhood of human habitations and remained there. When one knows how shy wild geese are, one can imagine the power of the herd instinct. Similar behavior occurs in a great many social vertebrates, up to chimpanzees, of which Yerkes rightly said, "One chimpanzee is no chimpanzee."

Even those instinct movements which, phylogenetically speaking, have quite recently achieved independence through ritualization, and which are the younger members in the parliament of instincts, can under certain conditions "shout down" all opposing drives such as hunger and sex. In the triumph ceremony of geese we shall learn about a ceremony that rules the life of these birds more than any other drive. On the other hand, there are of course numerous ritualized movement patterns which have barely made themselves independent of their unritualized prototypes and whose modest influence on general behavior consists only in the fact that the "desired" coordination of movements (as we have seen in the inciting movements of the Ruddy Shelduck; page 59) is slightly preferred and is more often performed than other possible forms of movements.

Whether a ritualized behavior pattern has a "strong" or "weak" voice in the concert of drives, it renders every motivation analysis difficult, because it can simulate a behavior arising from several independent drives. In the preceding chapter I have already said (page 62) that the ritualized movement, consisting of several components welded to a unit, copies the form of a movement pattern not determined by one hereditary co-ordination but originating in the conflict of several drives, as illustrated by the inciting of the duck. Since copy and original mostly overlap in the same movement, it is extremely difficult to analyze how much of it is caused by the copy and how much by the original. The new independent variable becomes clearly recognizable only if one of the originally independent components (in the inciting ceremony, the orientation toward the threatened enemy) comes in conflict with the ritualized co-ordination.

The zigzag dance of the male stickleback, on which Jan van Iersel conducted the first of all experimental motivation analyses, offers a good example of how a very weak ritual can creep as a hardly noticeable third, independent variable into the conflict of two "big" drives. Van Iersel noticed that the remarkable zigzag dance which is performed by a sexually mature, territory-owning male stickleback in front of every approaching female, and which was formerly explained simply as "courtship," looks different in every case. Sometimes it is the zig toward the female and sometimes it is the zag away from her which is the more stressed. If it is the zag, it is clear that this is directed toward the nest. In one extreme case, the male, seeing a female approaching, swims quickly toward her, brakes just in front of her, turns around—particularly when she immediately presents her fat belly to him—and swims back to the nest entrance; there, lying flat on his side, he shows his colorful flank and his bright green eye to the female in a special ceremony.

From these observations, van Iersel rightly concluded that the zig toward the female was activated by aggression, and the zag toward the nest by the sexual drive. He was able to demonstrate the correctness of his supposition, by his invention of methods by which he could exactly measure the strength of the aggressive drive and that of the sexual drive in a certain male. He offered the male a dummy rival of standardized size, and registered the intensity and duration of the fight reaction. He measured the sexual drive by showing the male an artificial female and, after a certain time, suddenly removing it. In such a case the stickleback male "discharged" the suddenly blocked sexual drive by a brood-tending action which normally serves to fan fresh water for the eggs or young into the nest. The duration of this displacement activity gives a reliable measure of the sexual motivation.

From these measurements van Iersel could now accurately predict what the zigzag dance of this particular male would look like and, conversely, from direct observation of the form of the dance he could estimate the relative participation of the two drives and the results of their subsequent measurements. The expert in ritualized movement patterns, seeing the rhythmical regularity with which the male stickleback alternates between zig and zag, will suspect that in this movement pattern, in addition to the two components determining its form, a third, if weaker one, is concerned. An alternation between the dominance of two opposing impulses hardly ever produces such regular oscillations unless a new motor coordination, formed by ritualization, is involved. Without this, little thrusts in different directions of space follow in very typical, irregular distribution, as we all know from the behavior of human beings in situations of extreme indecision. Ritualized movement, on the other hand, under the selection pressure of its function as an unambiguous signal, always tends to develop rhythmical repetitions of identical motor elements (page 76).

The suspicion that ritualization may be involved becomes a certainty when we see how the dancing male stickleback, during the zag, occasionally seems to forget completely that the dance, being motivated by the sexual drive, ought to point directly toward the nest, and now he describes a wonderfully regular, jagged circle around the female, in which all the zigs are directed toward her and all the zags away from her. In spite of the relative weakness of the new motor co-ordination which strives to turn the zig and the zag into a rhythmical zig-zag, it can sway the balance between the two motivations and effect the regular alternation of their motoric actions. The second important effect by which a ritualized co-ordination can become apparent is the alteration in spatial direction of the basic, unritualized movement produced by other impulses. Of this we already have an example in the classical prototype of a rite, namely the inciting ceremony of the Mallard duck (page 60).

Chapter Seven

Behavioral Analogies to Morality

In Chapter Five, on ritualization, I have tried to show how this process, whose causation is still so mysterious, creates new instincts that dictate to the organism their own "Thou shalt," as irresistible as any of the allegedly irresistible drives of hunger, fear, or sex. In the previous chapter, Six, I have attempted the still more difficult task of explaining how the play of multiple interactions between the different autonomous instincts determines behavior. I have tried to show what general rules it obeys, and what methods we can use to help us, in spite of all complications, to gain some insight into the causation of those behavior patterns that arise from several conflicting drives.

I hope I have succeeded so far that I can now proceed to synthesize the results of the last two chapters and to draw some conclusions relevant to the question concerning us here: How does the rite accomplish the seemingly impossible task of preventing those effects of intra-specific aggression which are injurious to communal life, without at the same time eliminating those of its functions that are essential to the survival of the species? The stipulation expressed in the last words of the last sentence is really the answer to a question which readily suggests itself but which quite mistakes the nature of aggression. This question is: Why has aggression not simply been eliminated in those animal species in which close social aggregation is of advantage to survival? The reason is that the functions dealt with in Chapter Three are indispensable.

The problem thus presented to the two great constructors of evolution is always solved in the same manner: the generally useful, indispensable drive remains unaltered, but for the particular case in which it might prove harmful, a very special inhibitive mechanism is constructed *ad hoc*. Here again the cultural evolution of human peoples proceeds analogously, and that is the reason why the most important imperatives of the Mosaic, as of all other laws, are not commandments but prohibitions. Later on we shall have to discuss the fact that it is only the rare, inspired maker of law who, in creating a taboo, is acting on the principle of conscious, rational morality. The devout and orthodox followers of the lawmaker obey his commandments for the nonrational reasons discussed in Chapter Five. Like the instinctive inhibitions and rites which prevent antisocial behavior in animals, the taboo effects a motivation which is analogous to true morality in function only, and which in all other respects is as far beneath it as the acquiring of conditioned response is beneath conceptual, rational thought; in other words, as far as the animal is beneath humanity. However, nobody with a real appreciation of the phenomena under discussion can fail to have an ever recurring sense of admiration for those physiological mechanisms which enforce, in animals, selfless behavior aimed toward the good of the community, and which work in the same way as the moral law in human beings.

An impressive example of behavior analogous to human morality can be seen in the ritualized fighting of many vertebrates. Its whole organization aims at fulfilling the most important function of the rival fight, namely to ascertain which partner is stronger, without hurting the weaker. Since all human sport has a similar aim, ritualized fights give the impression of "chivalry" or "sporting fairness." To this quality a cichlid species, *Cichlasoma biocellatum,* owes its American nickname. It is called "Jack Dempsey," after the world champion boxer renowned for the fairness of his fighting.

We know a good deal about the ritualized fights of fish and about the processes of ritualization by which they evolved from the original damaging modes of fighting. In nearly all teleosts, the fight is preceded by threatening movements which, as we have already described, always arise from the conflict between aggression and escape drive. Of these movements, the so-called broadside display has developed into a special rite which primarily arose through a fear-motivated turning away from the opponent, and a simultaneous escape-motivated spreading of the vertical fins. These movements have the result of presenting, to the adversary, the largest possible contours of the fish, making it appear bigger and more fear-inspiring. This desirable effect exerted a selection pressure which, in very many groups of fish, caused the evolution of exaggerated threat gestures in which expanded fins are displayed broadside-on. It is in the service of this broadside display that, in cichlids, in the Siamese fighting fish, and many others, the vertical fins have attained the beautiful development of size, form, and color which have made these fish so popular with aquarists.

In close connection with broadside threatening, in teleosts, the widespread intimidation gesture, the so-called tail beat, has arisen. From the broadside position, the fish, with stiffly held body and widely spread tail fin, makes a strong tail stroke at its opponent. The opponent is never touched, but receives by way of the pressure sense organ in its side a pressure wave whose strength evidently informs it, in the same way as does the extent of the visible body contours, of the size and fighting power of its enemy.

In many perchlike fishes, but also in other teleosts, another form of threatening has arisen from the ritualization of a frontal attack inhibited by fear. Each of the two adversaries swims straight at the other, preparing but not quite daring to deliver a ramming thrust. Their bodies tense and twisted like S-shaped springs, the opponents swim slowly toward each other and come to a standstill head to head, usually spreading the gill covers and blowing out the branchial membrane, thus enlarging the body contours visible to the enemy. In many fish, following frontal aggression, both opponents sometimes snap simultaneously at the presented mouth of the other. Depending on the conflict situation from which frontal threatening arises, they do so not resolutely but rather hesitatingly. In some fish families, for example in Labyrinth fishes, which are only loosely related to the Perchlike fishes, among which the Cichlids also belong, a highly interesting ritualized fighting method has evolved from this form of mouth fight, in which both rivals literally "match their strength" without injuring each other. They seize each other by the jaws and pull with all their might. In all species in which such ritualized mouth fighting occurs, lips and jaws are covered with a thick invulnerable leather skin so that the ensuing wrestling match is quite harmless. It is strongly reminiscent of the old Swiss farmers' sport of *Hosenwrangeln,* in that endurance rather than momentary strength decides the issue. When the opponents are equally matched the contest can last literally for hours, and in the case of two males of equal strength we recorded a ring fight of this type lasting from 8:30 A.M. to 2:30 P.M.

This so-called "mouth fighting"—in some species the fishes push, in others they pull each other—is followed after a certain length of time, varying from one species to another, by the primitive injury-inflicting way of fighting in which the fish try to ram each other in the unprotected flank with no holds barred, and to wound each other seriously. The injury-preventing rites of threatening and the subsequent measuring of strength represent, in their original form, only an elaborate introduction to the real, ruthless battle. But such a prolonged introduction fulfills an extremely important function, in that it enables the weaker rival to withdraw in time from a hopeless contest. Thus in most cases the species-preserving function of the rival fight, selection of the stronger, is fulfilled without the loss or even wounding of one of the individuals. It is only in the rare cases in which the fighters are of exactly equal strength that a decision can be reached only by bloodshed.

The comparison between species with less and those with more highly developed ritualized fights, and the study of the developmental stages leading, in the life of the individual, from the young fish fighting without rules to the "fair" Jack Dempsey, give us definite clues as to how ritual fights have evolved. There are three separate processes leading to the evolution of the chivalrous ritual fight from the "catch-as-catch-can" of the injury-inflicting fight, and ritualization is only one of these, though it is certainly the most important of the three.

The first step from the injury-inflicting to the ritualized fight consists of the lengthening of the periods between the single, gradually increasing threatening movements and the final assault. In purely injury-inflicting species, such as the Multicolored Mouthbreeder, the single phases of threatening, fin spreading, broadside display, blowing out the branchial membrane, tail beating, and mouth fight, last only seconds before the first wounding blows are dealt in the flank of the opponent. In the quick rise and fall of excitation so characteristic of these irascible fishes, single stages of threatening are sometimes passed over, and a particularly "quick-tempered" male may come to the point so quickly that he opens hostilities immediately with serious ramming. This is never seen in the closely related African species of Jewel Fish, which always maintains the right sequence of threatening movements and performs each of these for a longer time—often many minutes—before going on to the next one.

For this purely temporal division of movement there are two possible physiological explanations: one is that the threshold values of excitation, to which the single movement patterns respond as the fighting spirit mounts, are moved further apart so that their sequential order is preserved even when excitation rises and falls. The other possible explanation is that the increase of excitation is throttled and forced into a smooth and regular ascendancy curve. Reasons too complicated to deal with here speak for the first of these two hypotheses.

The longer they last, the more and more ritualized the single threatening movements become, and this leads, as already described, to mimic exaggeration, rhythmical repetition, and to the rise of structures and colors which accentuate the movement optically. Enlarged fins with color patterns visible only on spreading, ostentatious eye-spots on the gill cover or on the branchial membrane, strikingly displayed during frontal threatening, and various other theatrical effects make the ritualized fight one of the most attractive spectacles to be seen in the study of the behavior of higher animals. The colors glowing with excitement, the measured rhythm of the threatening movements, the exuberant strength of the rivals, make us almost forget that this is a real fight and not an artistic display performed for its own sake.

The third process contributing toward the transformation of the dangerous injury-inflicting fight into the noble ritualized one is just as important for our main theme as is ritualization. Special physiological mechanisms have evolved to inhibit the injury-inflicting movements. For example, when two

"Jack Dempseys" have opposed each other long enough with broadside threatening and tail beating, one of them may be inclined to go on to mouth pulling a few seconds before the other one. He now turns from the broadside position and thrusts with open jaws at his rival, who, however, continues his broadside threatening, so that his unprotected flank is presented to the teeth of his enemy. But the aggressor never takes advantage of this; he always stops his thrust before his teeth have touched the skin of his adversary.

My friend the late Horst Siewert described and filmed an analogous process among Fallow Deer. In these animals, the highly ritualized antler fight, in which the crowns are swung into collision, locked together, and then swung to and fro in a special manner, is preceded by a broadside display in which both animals goose-step beside each other, at the same time nodding their heads to make the great antlers wave up and down. Suddenly, as if in obedience to an order, both stand still, swing at right angles toward each other, and lower their heads so that their antlers collide with a crash and entangle near the ground. A harmless wrestling match follows, in which, just as in the mouth fights of Jack Dempseys, the victor is the one with the longest endurance span. Among Fallow Deer, too, one of the fighters sometimes wants to proceed, in advance of the other, to the second stage of the fight and thus finds his weapon aimed at the unprotected flank of his rival—a highly alarming spectacle, considering the formidable thrust of the heavy, jagged antlers. But more quickly even than the cichlid the deer stops the movement, raises his head, and now, seeing that his unwitting, still goose-stepping enemy is already several yards ahead, breaks into a trot till he has caught up with him and walks calmly, antlers nodding, in goose-step beside him, till the next thrust of the antlers leads, in better synchronization, to the ring fight.

Among the higher vertebrates, there are countless examples

of such inhibitions against injuring fellow members of the species, and they often play an essential part in situations where the anthropomorphizing observer would never suspect that aggression was present or that special mechanisms were necessary for its suppression. For example, to people who believe in the "infallibility" of instinct it will seem paradoxical that an animal mother has to be prevented, by special inhibitions, from aggressiveness toward her own children, particularly toward the new-born or newly hatched.

In reality, these special inhibitions against aggression are very necessary because a brood-tending animal parent, at the time when it has young, must be particularly aggressive toward every other living creature. In defense of her brood, a mother bird must attack every living creature that approaches her nest, provided that she is more or less a match for it. As long as she is sitting on her nest, a turkey hen must constantly be prepared to attack with all her might mice, rats, weasels, crows, magpies, etc., also members of her own species, a cock or a nest-seeking hen, which are just as dangerous to her brood as enemies wanting to devour them. She must be the more aggressive the nearer the threat to the center of her world, that is, to her nest. The only creatures she must not harm are her own chicks, which hatch from the egg just as her aggression reaches boiling point. My research associates, Wolfgang and Margaret Schleidt, discovered that this inhibition in the turkey hen is elicited acoustically only. For the examination of certain reactions of the turkey cock to acoustic stimuli, a number of poults were rendered deaf by an operation on the inner ear. Since this can be done only on freshly hatched chicks in which the sexes are difficult to differentiate, there were some unwanted females among them and these, being of no use for anything else, were used in testing the function of response behavior which plays such an essential part in the relation between mother and child. We know, for example, that freshly hatched greylag goslings regard as mother that object which responds with sound expressions to their "distress cheeping." The Schleidts wanted to make freshly hatched turkey poults choose between a hen which could hear them and which would therefore answer their cheeping in the right way, and a deaf one which they expected would call haphazardly without relation to the cheeping of the poults.

As so often in behavior research, the experiment demonstrated something which nobody was expecting, but which was much more interesting than the expected results. The deaf turkey hens incubated quite normally, and previously to this their social and mating behavior was likewise normal, but when their chicks hatched their maternal behavior proved to be affected in a highly dramatic way: all the deaf hens pecked all their children to death as soon as they hatched. A deaf hen which has sat on false eggs for the normal incubating period should be prepared to accept chicks, but if she is shown a day-old poult she does not react with maternal behavior: she utters no call notes but, if the baby approaches within yards, she raises her feathers defensively, hisses furiously, and as soon as the chick is within reach of her beak she pecks it as hard as she can. If we assume that the hen is in no way deranged other than in her hearing faculty, there can be only one interpretation of this behavior: she does not possess the slightest innate information as to what her chicks should look like, and she pecks at everything which moves near her nest and which is not so big that her escape reaction transcends her aggression. Only the sound expression of the cheeping chick elicits innate maternal behavior and puts aggression under inhibition.

The following experiments with normally hearing turkey hens confirmed the accuracy of this interpretation. If a natural-looking, stuffed chick is pulled, like a puppet, on a long string toward a brooding turkey hen, she pecks at it exactly as the deaf hen does. But if we build into the dummy a little loud-speaker which transmits from a tape recording the "crying" of a turkey poult, the intervention of an evidently violent inhibition will brake the attack just as suddenly as in the case of the cichlids described above, and the hen will begin to utter the typical call notes which, in the turkey, correspond to the clucking of the barnyard hen.

Every inexperienced turkey hen, brooding for the first time, attacks all objects which move near her nest and which are of a size approximately between that of a mouse and that of a large cat. The bird does not know innately what predators look like; she pecks at a dumb weasel or golden hamster no more fiercely than at a stuffed turkey poult. Conversely, she is immediately prepared to treat the first two maternally, if they can "identify" themselves by a built-in loud-speaker. It is impressive to observe how a turkey hen which has just been pecking furiously at an approaching dumb chick will spread herself out, to the accompaniment of motherly calling, to allow a cheeping polecat to creep under her.

The only characteristic which seems to strengthen the innate reaction to the nest enemy is a hairy, furry superficial consistency; at least, it seemed to us, in our first experiments, as though fur dummies had a stronger effect than smooth ones. Since a turkey poult has the right size, moves about near the nest, and in addition wears a downy coat, it cannot avoid constantly eliciting in the mother behavior patterns of nest defense which must be equally constantly suppressed by chick sounds if infanticide is to be prevented. This certainly applies to hens brooding for the first time and having no experience of the appearance of their own children. However, individual learning alters these behavior patterns very quickly.

The "maternal behavior" of the turkey hen, so full of contradictions, as I have shown, gives us food for thought. Something which can be described, as a whole, as "maternal instinct" or "brood-tending instinct" evidently does not exist, nor does an innate pattern, an innate recognition of the animal's own young. The species-preserving behavior toward the young is the function of a number of phylogenetic behavior patterns, reactions, and inhibitions so organized by the great constructors that, under normal environmental conditions, they co-operate as a systemic whole, "as though" the particular animal knew what it had to do in the interests of survival of the species. This system is what is commonly known as "instinct": in the case of our turkey hen, "brood-tending instinct." However, this concept, even when interpreted as above, is misleading in that it is not a demarcated system fulfilling the concept-determining functions. To be more precise, drives are built into its organization, which have quite other functions, as for example aggression and aggression-eliciting mechanisms. The fact that the turkey hen is infuriated by the sight of a fluffy chicken running about near her nest is not an undersirable side effect, but it is favorable to brood defense if the mother is put by her children into a state of irritability and aggressiveness. She is prevented from attacking them by the inhibition elicited by their cheeping, and so she comes to vent her anger on other creatures that approach her nest. The only specific organization that functions in this one behavior system only is the selective response of the pecking inhibition to the chicks' cheeping.

The fact that animal mothers of brood-tending species do not attack their young is thus in no way a self-evident law, but has to be ensured in every single species by a special inhibition such as the one we have learned about in the turkey hen. Every livestock breeder knows what apparently slight disturbances can cause the failure of an inhibition mechanism of this kind. I know of a case where an airplane, flying low over a silver-fox farm, caused all the mother vixens to eat their young.

In many vertebrates which do not tend their young and in several which do so for a limited time only, the young at an early age, often before attaining full size, are relatively as strong and, since such species are not capable of much learning, as clever as the adults. Thus they do not stand in need of particular protection, nor are they treated with any consideration by the adults. However, the situation is quite different in those most highly organized animals in which learning and individual experience play a big part and in which parental care is prolonged, because the "school of life" of the children lasts a long time. Several biologists and sociologists have called attention to the intimate connection between learning capacity and duration of parental care.

A young dog, wolf, or crow after reaching its full body length—if not its full body weight—is still a clumsy, loose-limbed animal neither able to defend itself against a serious attack by an adult of its species nor able to escape from it by rapid flight. One would think that the ability to escape quickly would be particularly necessary in these species and in many other similar ones, since the young are defenseless not only against intra-specific aggression but also against the predatory habits of adults of their own species. However, cannibalism is very rare in warm-blooded vertebrates and almost unknown in mammals, probably for the simple reason that conspecifics "do not taste good," a fact observed by polar research scientists when they tried feeding the flesh of dead or emergency-slaughtered dogs to the team survivors. Only some birds of prey, particularly goshawks, will kill and eat their fellows in captivity, but I know of no case where this has been observed in the wild state. It is not yet known what inhibitions prevent it.

For the full-grown but still clumsy young animal, a much greater danger than cannibalism lies in the aggressive behavior of the adults. This danger is precluded by a series of strictly regulated inhibitory mechanisms, as yet largely unexplained. In one case, however, these behavior mechanisms are easily analyzed: this is in the loveless society of night herons, to which we shall devote a special chapter. The young birds are able to remain in the colony, in spite of the fact that within its narrow confines every tree branch is the subject of jealous disputes between territorial neighbors. As long as the young, fledged heron continues to beg, it is sure of absolute protection against all attacks of territorial adult herons. Before an older bird can even get ready to peck at a young one, the young bird importunes it with begging calls and wing flapping, and tries to seize its beak, pulling it down and "milking" it, as heron babies do to the beak of their parents when they want regurgitated food. The young night heron does not know its parents personally, and I am not sure that the parents know their children either; only the baby birds in the nest certainly know each other. Just as the adult night heron, when not in the mood to feed babies, flees before the clamoring of its own children, so it flees before every strange young bird and does not dream of attacking it. We know of analogous cases among many animals in which infantile behavior protects against intra-specific aggression.

A still simpler mechanism enables the young Night Heron, which is independent but not yet a match for its elders, to acquire a small territory inside the colony boundaries. The striped juvenile plumage, worn by the young bird for nearly three years, elicits far less intensive aggression in the adult than the finished nuptial plumage does. A young heron lands somewhat aimlessly somewhere in the heronry and has the luck not to alight in the fiercely defended territorial center, that is, in the immediate vicinity of the nest of a breeding bird. Nevertheless an adult bird, feeling provoked, starts moving in the characteristic night heron way, slinking slowly and threateningly toward the newcomer; in so doing it inevitably comes too near the territory of another breeding bird. Since its plumage and threatening attitude are far more strongly aggression-releasing than the motionless, frightened young heron, the neighbors regularly make the elder bird the target of their counterattack, passing by the young one and thus involuntarily protecting it. Therefore birds in juvenile plumage are regularly seen to settle between the territories of resident breeding birds, in sharply circumscribed places where a heron in nuptial plumage would elicit the attack of all neighboring territory owners.

It is less easy to understand the inhibitive mechanism preventing adult dogs of all European breeds from seriously attacking puppies under the age of seven to eight months. In Greenland Eskimo dogs, this inhibition, as Tinbergen observed, protects only the young dogs of the same pack, and there is no inhibition against biting strange puppies. Possibly this also applies to wolves. By what means the youthfulness of an animal is recognized by members of its species is not altogether clear. Size certainly has nothing to do with it, and a tiny, old, bad-tempered fox terrier is just as friendly and attack-inhibited toward a huge, clumsy, importunate St. Bernard baby as he is toward a puppy of his own breed. Probably the essential characters activating this inhibition lie in the behavior of the young dog, possibly also in its smell, judging by the way a puppy invites an adult to make a smell test. As soon as the approach of the adult seems at all dangerous, the puppy throws itself on its back, presenting its still naked baby belly and passing a few drops of urine which are promptly sniffed by the adult.

Even more interesting and more problematical than the inhibitions protecting full-grown but still clumsy young animals are those aggression-inhibiting behavior mechanisms that prevent "unchivalrous" behavior toward the "weaker sex." In Dancing Flies, whose behavior has already been described on page 65, in the Praying Mantis and many other insects, also in many spiders, the females are the stronger sex and special behavior mechanisms are necessary to prevent the happy bridegroom from being eaten too soon. In Mantids, the female often eats with relish the front half of the male while his hind half completes undauntedly the great work of reproduction.

However, we are less concerned with such bizarre phenomena than with the inhibitions which among so many birds and mammals, including man, hinder if they do not entirely prevent the ill-treatment of females. "You cannot hit a woman" is, as far as it concerns human beings, a maxim of only conditional validity. But in animals there is a whole series of species in which, under normal, that is, nonpathological, conditions, a male never seriously attacks a female.

This is true of our domestic dogs and doubtless of the wolf too. I would not trust a dog that bit bitches and would warn his owner to be most careful, especially if there were children in the house, for something must obviously be out of order with the social inhibitions of such an animal. When I once tried to mate my bitch Stasi, a Chow-Alsatian hybrid, to a huge Siberian wolf, she became jealous because I played with him and she attacked him in real earnest. He did nothing, except present his big gray shoulder to the snapping red fury, in order to receive her bites in a less vulnerable place. Similar absolute inhibitions against biting a female are found in hamsters, in certain finches, such as Goldfinches, and even in several reptiles, for example the South European Emerald Lizard.

In the male of this species, aggressive behavior patterns are elicited by the gorgeous colors of the rival, particularly by his glorious ultramarine throat and the green of the rest of his body which gives him his name. On the other hand, the female-biting inhibition is evidently dependent on smell characteristics, as G. Kitzler and I found out when, by means of crayons, we deceitfully applied male coloring to the female of our largest pair of these lizards. When we put her back in the large enclosure, the female, naturally unconscious of her appearance, ran toward the territory of her male, who threw himself upon the apparently male intruder, opening wide his jaws with the intention of biting. Then he perceived the female smell of the painted lady and checked his attack so suddenly he turned a somersault over her. Then he examined her carefully with his tongue and took no more notice of the fight-eliciting colors—a considerable achievement for a reptile! But the interesting fact was that for a long time after this evidently impressive experience, this chivalrous lizard examined real males with his tongue, that is to say, he checked upon their smell before attacking them. Apparently it had affected him very deeply that he had once nearly bitten a lady!

One would imagine that the females of those species whose males have an absolute inhibition against biting females would behave very offensively toward the whole male sex, but oddly enough the very reverse is the case. An aggressive, large female emerald lizard, which fights members of her own sex relentlessly, literally falls on her belly before the youngest, weakest male, even when he is scarcely a third as heavy as herself and when his masculinity has only just become recognizable by the blue shades on his throat, comparable to the first soft hairs on the face of a schoolboy. Female emerald lizards lift their fore-paws from the ground and move them rapidly up and down in a peculiar way as though they were playing the piano. This is the submissive gesture, common to all lizards of the genus Lacerta. Bitches, too, particularly in those breeds nearest the Nordic Wolf, show toward the dog of their choice, although he has never bitten them or given any other sign of his superiority, a submissive adoration akin to that which they show toward their human master. But most remarkable and inscrutable is the ranking order relation between female and male in many finches of the Carduelis family, to which the Siskin, the Goldfinch, the Bullfinch, the

Greenfinch, and many others, including the Canary, belong.

According to R. Hinde, during the reproductive season the female Greenfinch is superior to the male but during the rest of the year the male is superior to the female. One arrives at this conclusion by simply watching who pecks whom, and who evades the pecking of the other. In the Bullfinch, specially well known to us through the studies of J. Nicolai, we could conclude as a result of similar observations and inferences that the female of this species, in which the pairs remain together year in, year out, is once and for all time superior to the male in ranking order. She is always a little aggressive, sometimes pecks at her mate, and even in the greeting ceremony—the so-called "beak flirtation"—there is a considerable measure of aggression, though in a strictly ritual form. The male, however, never pecks his wife and, if we judge the ranking order of the partners exclusively by recording who is doing the pecking and who is being pecked, we must infer that she is plainly his superior. However, on looking more closely we come to the opposite conclusion. When a male bullfinch is pecked by his wife, he in no way assumes the submissive attitude but, on the contrary, he shows sexual self-display and tenderness. Thus he is not pushed by the pecking of his wife into a subordinate position, but, on the contrary, his passive behavior, the manner in which he accepts his wife's attacks without becoming aggressive and without letting himself be put out of a sexual mood, has an "impressive" effect—apparently not only on the human observer.

Analogous with the behavior of male bullfinches is that of a male dog and a male wolf toward all female attacks. Even when these are meant seriously, as in the case of my Stasi, ritual demands of the male not only that he should not bite back but that he should preserve his "friendly face," with ears laid back high on the head, and the skin of the forehead drawn smooth across the temples. Keep smiling! The only defense movement which I have seen in such cases and which is also mentioned by Jack London in *White Fang* consists of sideways catapulting with the hind-quarters; this has a "casting away" effect, particularly when a heavy dog, without losing his friendly smile, flings a snarling bitch some yards to one side.

We are not crediting the canine and bullfinch ladies with all too human qualities when we say that they are impressed by the passive acceptance of their aggression. That not-being-impressed makes a deep impression is a very general principle, as was shown also by an observation repeatedly made on fighting fence lizard males by G. Kitzler. In the wonderful ritual fights of the Fence Lizard, each of the rivals first holds his heavily armored head in an attitude of self-display toward the other, until one seizes the other. After a short wrestling match, he lets go and waits till the other in his turn seizes him. With rivals of equal strength, many such bouts take place until one of the contestants, unhurt but exhausted, gives up the fight. Now in lizards, as in many other "cold-blooded" animals, smaller individuals "get going" quicker than larger ones, that is the surge of a new excitation rises in them more quickly than in bigger and older members of their species. This means that in the ritual fight of the Fence Lizard, with a certain degree of regularity the smaller of the two fighters is the first to seize the other by the back of the head and to pull him to and fro. When there is much difference in size, it may happen that the smaller one, having let go, does not await the return bite of the larger one, but at once performs the submissive gesture and then flees. He has noticed from the purely passive resistance that his adversary is superior.

This procedure looks so funny that it always reminds me of a scene in a long-forgotten Charlie Chaplin film. Charlie creeps up behind his enormous rival with a huge piece of wood, swings it, and hits him with all his might on the back of the head; the giant looks up absent-mindedly and passes his hand several times over the place, obviously thinking that a fly has settled there. Thereupon Charlie turns and runs as only Charlie can.

In pigeons, songbirds, and parrots, there is a most remarkable ritual which bears a mysterious relation to the ranking order of the mates: feeding the mate. This ritual, usually considered by superficial observers to be "billing and cooing"—a form of kissing—is, like so many other apparently "selfless" and "chivalrous" behavior patterns in animals and men, not only a social duty but at the same time a privilege that falls to the lot of the higher-ranking individual. In fact, each of the two mates prefers feeding to being fed, on the principle, It is more blessed to give than to receive, or—if the crop is full to bursting—it is more blessed to get rid of food than to accept it. Under favorable circumstances it can be seen that a slight ranking order quarrel between the two mates is necessary in order to decide who may feed and who must play the less desirable part of the child, opening its beak and letting itself be fed.

Nicolai once reunited a pair of the small African Gray Serins after a long separation; the partners recognized each other at once and flew joyfully toward each other, but the female had evidently forgotten her former ranking relation to her partner, for she started to regurgitate food from her crop and tried to feed him. Since he did the same thing, a small difference of opinion arose, which was won by the male, whereupon the female no longer tried to feed, but submitted to being fed. Among bullfinches, where the partners stay together for the whole year, the male sometimes starts to molt before the female and thus reaches a low ebb of sexual and social aspiration, while the female is still well up to form in both respects. In such a case, which may occur under natural conditions, as well as in the rarer case of the male's losing precedence for pathological reasons, the normal direction of feeding is reversed and the female feeds the weakened male. The anthropomorphizing observer is moved by the fact that the female looks after her sickly mate, but after what we have said, this interpretation is wrong. She would always have fed him if she had not been prevented from doing so by her inferior status.

The social precedence of the female among bullfinches and canines is thus only an apparent one, and it is elicited by the "chivalrous" inhibition of the males against biting females. Western civilization offers a cultural analogy between human customs and animal ritualization, of exactly similar form. Even in America, the land of boundless respect for woman, a really submissive man is not appreciated. It is expected of the ideal male that, in spite of mental and physical superiority, he should submit according to ritually laid-down laws to the smallest whim of his wife; but there is an expression, taken from animal behavior, for the contemptible, really submissive man: he is called henpecked, a metaphor that well illustrates the abnormality of male submissiveness, for a real cock does not let himself be pecked by any hen, not even his favorite. Incidentally, cocks lack the inhibition against fighting hens.

The strongest inhibition against biting females of the species is found in the European Hamster, and the significance of this may be that the male is several times heavier than the female and his long incisors can inflict severe wounds. For the short mating season, the male enters the territory of the female and, as Eibl-Eibesfeldt observed, it is some time before the two recluses have become well enough acquainted for the female to tolerate the advances of the male. During this period, and only then, does the female hamster show fear and shyness of the male. At all other times she is like a fury, biting at him unrestrainedly. When these animals are bred in captivity, they have to be separated when mating is over, otherwise there would soon be male corpses.

Three features, mentioned above in the behavior pattern of the hamster, are peculiar to all mechanisms inhibiting killing and injuring: first, the correlation between the effectiveness of the weapons of an animal species, and the inhibition preventing their use against other members of the species; secondly, the special appeasing rites which aim at releasing these inhibitive mechanisms in aggressive members; thirdly, the important fact that there is no absolute reliance on these inhibitions, which may occasionally fail.

In a previous book, *King Solomon's Ring,* I have shown that those inhibitions which prevent animals from injuring or even killing fellow members of the species have to be strongest and most reliable, first in those species which being hunters of large prey possess weapons which could as easily kill a conspecific; and secondly, in those species which live gregariously. In solitary carnivores, as for example some marten and cat species, it suffices if sexual excitement effects a temporary inhibition of aggression and of preying, lasting long enough to enable the sexes to mate without danger. Large predators, however, which live permanently in a society as wolves or lions do, must possess reliable and permanently effective inhibition mechanisms. These must be sufficiently self-reliant to be independent of the changing moods of the individual. And so we find the strangely moving paradox that the most bloodthirsty predators, particularly the Wolf, called by Dante the *bestia senza pace,* are among the animals with the most reliable killing inhibitions in the world. When my grandchildren play with other children of the same age, supervision by an adult is advisable, but I do not hesitate to leave them unsupervised in the company of our big Chow-Alsatian dogs whose hunting instincts are of the bloodthirstiest. The social inhibitions on which I rely are certainly not those which have been bred into the dog in the course of his domestication, but without any doubt they are the heritage of the Wolf, the *bestia senza pace.*

The combinations of stimuli which set social inhibition mechanisms in action evidently vary greatly from species to species. As we have seen, the female-biting inhibition of the male Emerald Lizard is dependent on chemical stimuli; the dog's inhibition against biting bitches is also chemically induced, whereas his indulgence toward puppies is evidently elicited by their behavior. Since the inhibition is an active process opposing a likewise active drive, checking or modifying it, it is quite correct to speak of releasing an inhibition process, just as we speak of the releasing of an instinctive movement. The multiform stimulus-sending apparatuses which in all higher animals serve for the releasing of response behavior are basically no different from those which release social inhibition. In both cases, the stimulus senders consist of ostentatious structures, bright colors, and ritualized behavior patterns, generally combinations of all three. The stimulus senders that release activities and those that release inhibitions are built on exactly the same principle; a good illustration of this is seen in the releasers of fighting in cranes, and of the baby-biting inhibition in some rails. In both cases, on the back of the bird's head a little tonsure or naked patch has evolved, under the skin of which is a richly branched vascular net, a so-called *corpus cavernosum.* In both cases, this organ becomes filled with blood, forming a prominent little red cap which is presented to a fellow member of the species. The functions of these releasers, which in the two bird groups have arisen entirely independently of one another, are diametrically opposed: in cranes, the signal means an aggressive mood and so elicits, according to the relative strength of the opponent, either counteraggression or escape mood. In the Water Rail and several related species, the organ and the behavior pattern are present in the chick only and serve exclusively to elicit a specific chick-biting inhibition in older members of the species. "By mistake," water rail chicks often present their red cap tragicomically to aggressors of other species also. A hand-reared chick of mine did this to ducklings, which naturally did not respond with inhibitions to this specific signal of the Water Rail, but pecked the little red head. Soft though a duckling's beak may be, I nevertheless had to separate the birds.

Ritualized behavior patterns eliciting in members of the same species an inhibition against aggression are called submissive or appeasing attitudes, the second being a better term because it leads less to the subjectivizing of animal behavior. The origins of this kind of ceremony are as multiform as those of all other ritualized expression movements. In our discussion of ritualization we have already learned how signals serving social communication can arise from conflict behavior, intention movements, etc. We have also seen what power the rite evolved in this manner can wield in the great parliament of instincts. All these facts need to be explained as a basis for the understanding of the evolution and the function of appeasement movements.

Curiously enough, appeasement gestures have evolved in a large variety of animals under the selection pressure exerted by behavior patterns releasing aggression. In trying to appease a member of its species, the animal does everything to avoid stimulating its aggression. A cichlid, for instance, elicits aggression in another by displaying its colors, unfolding its fins, or spreading its gill covers to exhibit its body contours as fully as possible, and by moving in strong jerks; if the same fish wishes to appease a superior opponent it does exactly the opposite: it grows pale, draws in its fins, displays the narrow side of its body, and moves slowly, stealthily, literally stealing away all aggression-eliciting stimuli. A cock beaten in a rival fight puts its head in a corner, thus removing from its opponent the fight-eliciting stimuli that come from the red comb and wattles. We have already learned that certain coral fish, whose bright coloring elicits intra-specific aggression, divest themselves of this coloring when they want to approach each other peacefully for mating.

The removal of the fight-eliciting signal only prevents the stimulation of intra-specific aggression. It does not, in itself, release an active inhibition against an attack already in progress. But considered phylogenetically, it is obviously only a step from the one to the other, and the origin of appeasement gestures from the "negative" fight-eliciting signal is a good example of this fact. In many animals, the threat consists in holding the weapon, whether it be teeth, beak, claws, wing, or fist, significantly "under the nose" of the opponent; and since all these gestures belong to the innately "understood" signals, which according to the relative strength of the threatened animal elicit counterattack or flight, the path for the origin of fight-hindering gestures is clearly marked: the suppliant must turn his weapon away from his opponent.

However, since the weapon serves not only for attack but also for defense, this form of appeasement gesture has the disadvantage that every animal which performs it renders itself defenseless, and, in many cases, additionally offers to the potential aggressor the most vulnerable parts of its body. Nevertheless, this form of submissive gesture is very widespread and has been "invented" by the most various vertebrates. The wolf turns his head away from his opponent, offering him the vulnerable, arched side of his neck; the jackdaw holds under the beak of the aggressor the unprotected base of the skull, the very place which these birds attack when they intend to kill. The connection is so obvious that for a long time I believed that the presentation of the most vulnerable part was essential for the effectiveness of submissive attitudes. In the wolf and the dog, it really looks as if the suppliant is offering his neck veins to the victor. Even if the turning away of the weapon is primarily the sole effective constituent of this particular expression movement, there is a certain measure of truth in my former opinion.

It would indeed be suicidal if an animal presented to an opponent still at the height of aggressiveness a very vulnerable part of its body, acting on the supposition that the simultaneous switching-off of fight-eliciting stimuli would suffice to prevent attack. We all know too well how slowly the balance changes from the predominance of one drive to that of another, and we can safely assert that a simple removal of fight-eliciting stimuli would effect only a very gradual ebbing of the aggressive mood. When a *sudden* presentation of the submissive attitude inhibits the threatened attack, we can safely assume that an active inhibition was elicited by a specific stimulus situation.

This is certainly the case in the dog, in which I have repeatedly seen that when the loser of a fight suddenly adopted the submissive attitude, and presented his unprotected neck, the winner performed the movement of shaking to death, in the air, close to the neck of the morally vanquished dog, but with closed mouth, that is, without biting. In Gulls, there is a similar behavior in the Kittiwake, and in corvide birds in the Jackdaw. Among the gulls whose behavior is well known to us through Tinbergen and his school, the Kittiwake holds a special place since, owing to an ecological peculiarity—the practice of nesting on narrow ledges of steep rock—the young have necessarily to stay in the nest until fledging. Thus the nestlings require better protection against possible attacks from strange gulls than do the nestlings of ground-breeding species which can run away if necessary. Correspondingly the appeasement attitude of the Kittiwake is not only more highly developed but it is enhanced by a special color pattern in the young bird. In all gulls, a turning away of the beak from the opponent acts as an appeasement gesture, but while in the Herring Gull and the Lesser Black-backed Gull, and in other large gulls of the genus Larus, this is not particularly noticeable and not suggestive of a special ritual, in the Black-headed Gull it is an exact, elaborate ceremony in which one partner turns the base of the skull toward the other. Occasionally, if neither of them harbors aggressive intentions, they do this simultaneously, twisting their heads through an angle of 180°. This "head flagging" is optically stressed by the fact that the black-brown mask and the dark red beak disappear suddenly while the snow-white neck feathers take their place. While in the Black-headed Gull the disappearance of the aggression-eliciting characters of the black mask and the red beak are still the effective components of the ceremony, in the young Kittiwake the presentation of the neck is particularly accentuated by the color pattern: on a white background there appears a characteristic dark marking which obviously effects a special inhibition of aggressive behavior.

A parallel to this evolution of an aggression-inhibiting signal in sea gulls is found in Corvidae, the family to which ravens, crows, etc., belong. All big black and gray species turn the head markedly away as an appeasement gesture. In many of these birds, the proffered nape region is emphasized by light coloring. In the Jackdaw, whose social life in the colony is a very close one, and which evidently requires a specially effective appeasing gesture, this head region is not only set off from the rest of the gray-black feathers by a silky pale gray coloring, but it has much longer feathers; these, like the ornamental plumes of several herons, lack the hooks on the barbules, so that they stand out like a shining crown when they are presented, maximally ruffled, to the beak of the aggressor. When this occurs the aggressor never attacks, even if on the verge of doing so, when the weaker bird assumes the submissive attitude. In most cases, the still angry aggressor reacts with the behavior of social grooming, preening and cleaning the back of the submissive bird's head, in quite a friendly manner—a really moving form of making peace!

There is another form of submissive gesture which derives from infantile behavior patterns, and there are still others which arise from the soliciting behavior of the female. In their present function, the gestures have nothing to do with infantility or with female sexuality, but they mean, in terms of human language, nothing more than "Please do not hurt me!" It is probable that in these particular animal species, before these expression movements achieved general social significance, special inhibitions prevented attack of the young or of females. It may also be assumed that in these species the bigger social group evolved from the pair and the family.

In our dogs, in the Wolf, and in other members of the same family, submissive or appeasing movements have evolved from juvenile expression movements persisting into adulthood. This is not surprising to anyone who knows how strong the inhibition against attacking pups is in any normal dog. R. Schenkel has shown that a great many gestures of active submission, that is of being submissive and, at the same time, *friendly* to a "respected" but not feared, higher-ranking animal, arise directly from the relation of the young animal to its mother: nuzzling, pawing, licking the corners of the mouth, movements which we all know well in friendly dogs, are derived, according to Schenkel, from movement patterns of sucking and begging for food. Just as two polite people may mutually express deference, though there is in reality a definite ranking order relationship between them, so two friendly dogs alternately perform infantile submissive gestures, particularly when greeting each other after a period of separation. This mutual politeness goes so far that, in his observations on free-living wolves on Mount McKinley, Murie was often unable to discover, from the expression movements of greeting, the ranking order relation of two adult male wolves. In the National Park on Isle Royale in Lake Superior, S. L. Allen and L. D. Mech observed an unexpected function of the greeting ceremony. In winter, the pack of about twenty wolves lives on moose and, according to observations, on weakened ones only. The wolves bring to bay every moose they can find, but they do not immediately attempt to savage it, and relinquish the attack if the victim puts up a strong defense. If, however, they find a moose that is debilitated by worms or illness or, as so often in old animals, by dental fistulae, they at once know that it is a suitable prey. Then all the members of the pack suddenly gather together and indulge in a ceremony of general nuzzling and tail wagging, movement patterns that we see in our dogs when we let them out of their kennels for exercise. This nose-to-nose conference signifies without any doubt the decision that the hunt is about to begin. Here we are reminded of Masai warriors who, in a ceremonial dance, work themselves into the necessary state of courage for a lion hunt.

Expression movements of social submissiveness, evolved from the female invitation to mate, are found in monkeys, particularly baboons. The ritual presentation of the hindquarters, which for purposes of visual emphasis are often incredibly colorful, has in its present form almost nothing to do with sexual motivation. It means that the individual performing the ritual acknowledges the higher rank of the one to whom it is directed. Even quite young baboons perform this ceremony without having been taught. When Katharina Heinroth's female baboon, which had lived with human beings since shortly after birth, was let into an unfamiliar room, she performed the ceremony of "presenting her behind" to every chair that apparently evoked her fear. A male baboon treats the females of his species somewhat brutally and dictatorially; according to observations by Washburn and de Vore he is not so brutal in the wild state as in confinement, but nevertheless his behavior is not gentle in comparison with the ceremonial politeness of male dogs, bullfinches, or greylags. So it is understandable that, in these monkeys, the two interpretations, "I am your woman" and "I am your slave," are more or less synonymous. The origin of this remarkable gesture expresses itself not only in the movement form but also in the way in which it is interpreted by the addressee. In the Berlin Zoo, I once watched two strong old male Hamadryas Baboons assaulting each other in real earnest for a minute. A moment later, one of them fled, hotly pursued by the other, who finally chased him into a corner. Unable to escape, the loser took refuge in the submissive gesture, whereupon the winner turned away and walked off, stiff-legged, in an attitude of self-display. Upon this, the loser ran after him and presented his hindquarters so persistently that the stronger one eventually "acknowledged" his submissiveness by mounting him with a bored expression and performing a few perfunctory copulatory movements. Only then was the submissive one apparently satisfied that his rebellion had been forgiven.

Of all the various appeasement ceremonies, with their many different roots, the most important for our theme are those appeasing or greeting rites which have arisen from redirected aggression movements. They differ from all the already described appeasement ceremonies in that they do not put aggression under inhibition but divert it from certain members of the species and canalize it in the direction of others. This new orientation of aggressive behavior is one of the most ingenious inventions of evolution, but it is even more than that: wherever redirected rituals of appeasement are observed, the ceremony is bound to the individuality of the participating partners. The aggression of a particular individual is diverted from a second, equally particular individual, while its discharge against all other, anonymous members of the species is not inhibited. Thus discrimination between friend and stranger arises, and for the first time in the world personal bonds between individuals come into being. If it is argued that animals are not persons, I must reply by saying that personality begins where, of two individuals, each one plays in the life of the other a part that cannot easily be played by any other member of the species. In other words, personality begins where personal bonds are formed for the first time.

As far as their origin and their original function are concerned, personal bonds belong to the aggression-inhibiting, appeasing behavior mechanisms, and therefore their place in this book should really have been in the present chapter on behavioral analogies to morality; but they form such an indispensable foundation for the building up of human society and are thus so important for our theme that they must be dealt with in a separate chapter. However, three other chapters must precede this, for only when we have learned of other possible societies in which personal friendship and love play no part, can we measure the full significance of these bonds in the human social structure. So I shall go on to describe the anonymous flock, the loveless society of Night Herons, and finally the society of rats, which inspires respect as well as repugnance, before I turn to the natural history of the strongest and most beautiful bond on earth.

Chapter Eight

Anonymity of the Flock

I am now going to take three forms of society as a primitive dark background with which to compare the society founded on personal friendship and love. The first of the three societies is the aggregation of anonymous members. It is the commonest and doubtless the most primitive form of animal association, and it is already found in many invertebrates such as Cephalopods, for example Cuttlefish and Squids, as well as in many insects. This, however, does not mean that it does not occur in higher animals; under certain horrible conditions even man can "regress" to anonymous herd formation.

By "flock" or "herd" we do not mean that chance gathering of like individuals such as occurs when many flies or vultures crowd around a carcass, or when many winkles or sea anemones settle on a particularly favorable place in the tidal zone. The concept of the flock is determined by the fact that individuals of a species react to each other by attraction and are held together by behavior patterns which one or more individuals elicit in the others. Thus it is typical of flock formation when many individuals travel in close formation in the same direction.

The questions confronting the behavior physiologist, who is

trying to understand flock formation, do not only concern the mechanisms causing the individual to seek the company of its own kind, but they also, more particularly, concern the high selectivity of these reactions. It calls for explanation when a herd animal wants at all costs to be near a lot of other members of its own species, and only in dire necessity will content itself with animals of other species as substitutes. This herding together may be innate, as for example in many ducks which react selectively to the signal of wing coloring in their own species by flying after it, or it may depend on individual learning.

We shall not be able to give a satisfactory answer to the many "Whys" regarding the herding together of anonymous crowds before we have solved the problem, "What for?"—in other words, before we have answered Darwin's question concerning survival value. In trying to do this, we meet with a paradox: it was easy to find a convincing answer to the apparently senseless question, "What is the good of aggression?" and we learned in Chapter Three of its species-preserving functions, but it is extremely difficult to say wherein lies the survival value of the aggregation of the huge anonymous herds which we find in fish, birds, and many mammals. We are too accustomed to seeing such communities, and since we ourselves are social beings we can appreciate that a herring, a starling, or a bison cannot feel happy by itself. And so it does not occur to us to ask, "What is it for?" However, the justification for this question will immediately be apparent if we consider the obvious disadvantages of big herds, for instance, the difficulty of finding enough food for so many animals, the impossibility of concealment, the increased predisposition to disease, and many other factors.

One would imagine that one herring swimming alone through the sea, one starling setting forth independently on its wanderings, or one lemming searching alone, in times of famine, for fertile fields, would have better chances of survival than would the dense crowds in which these animals herd together, and which provoke their own extermination by hunters and fishermen. We know that the drive forcing the animals together is a tremendously strong one, and that the attraction exercised by the herd over the individual or over smaller groups of individuals increases with the size of the herd, probably in geometrical proportion. Thus in many animals, for example bramblings, a deadly vicious circle may arise. Under the influence of fortuitous external conditions, such as a particularly good beechnut harvest in a certain area, the flocking together of these birds may far exceed its usual extent, the avalanchelike swelling of numbers may exceed the ecologically supportable limit, and the birds may starve in masses. In the winter of 1951 I had the opportunity of studying an enormous flock of these birds on the Swiss Thunersee. Every day there were many corpses under their roosting trees. Post-mortem examinations revealed that the birds had died of starvation.

I think we can conclude, from the proven disadvantages of life in big herds, that there must be advantages which not only compensate for but also so far outweigh the disadvantages that a selection pressure has arisen causing the evolution, among so many animals, of a complicated behavior mechanism of herding together.

If herd animals are in the smallest degree capable of defense against predators, as are jackdaws, small ungulates, and small monkeys, it is understandable that there should be safety in numbers. The repulsion of a predator, or the succor of an assaulted member of the herd, need not even be particularly effective in order to gain species-preserving value. The social defense reaction of jackdaws may not result in saving their fellow from a hawk, but if it is just annoying enough to make him hunt jackdaws a little less eagerly than magpies and thus to make him prefer magpies as prey, this is enough to give social defense a very strong survival value. The same applies to the alarm call with which a roebuck pursues predators, or to the malicious screeching to the accompaniment of which many small apes, from the safety of the treetops, go leaping after a tiger or a leopard. From such beginnings and by comprehensible transitions, the heavily armed defense organizations of bull buffaloes, baboon males, and similar heroes have evolved, from whose defensive power even the most terrible predators shrink.

But what advantage does close herding together bring to the completely defenseless, such as herrings and other small shoal fishes, small birds in enormous flocks, and many others? I can think of only one explanation and I offer it tentatively because, even to me, it seems scarcely believable that a single, small, but widespread weakness in predators could have wrought such far-reaching consequences in the behavior of their prey: this weakness lies in the fact that many, perhaps all, predators which pursue a single prey are incapable of concentrating on one target if, at the same time, many others are crossing their field of vision. Just try, yourself, to catch a single specimen from out of a cage full of birds. Even if you do not want a particular individual but intend to empty the whole cage, you will be astonished to find how hard you have to concentrate on a specific bird in order to catch one at all. You will also notice how incredibly difficult it is to concentrate on a certain bird and not allow yourself to be diverted by an apparently easier target. The bird that seems easier to catch is almost never caught, because you have not been following its movements in the immediately preceding seconds and therefore cannot anticipate its next movements. Moreover, it is astonishing how often we reach in the line of the resultant of two equally tempting attractive forces.

Many predators apparently do the same thing when offered

a number of targets at the same time. It has been experimentally demonstrated that, paradoxically, goldfish catch fewer water fleas when they are offered too many at once. Automatically radar-guided missiles behave in the same way when aiming at airplanes: they fly through the resultant between two targets if these are close together and positioned symmetrically on both sides of the projectile's trajectory. The predatory fish, like the radar-guided missile, lacks the ability to blind itself voluntarily to one objective in order to concentrate on the other. Probably herrings swim in close shoal formation for the same reason as jet fighters fly in close formation across the sky—a strategy not without danger even for expert airmen.

Far-fetched though this explanation of a widespread phenomenon may seem, strong arguments speak for its correctness. As far as I know, there is not a single gregarious animal species whose individuals do not press together when alarmed, that is, whenever there is a suspicion that a predator is close at hand. The smallest and most defenseless animals do this the most noticeably, and in many fish species only the small, young ones do it, while the adults do not. When in danger, some species of fish crowd together to form a body so that they look like one big fish, and, since many of the large, rather stupid predators such as the Barracuda meticulously avoid large prey for fear of choking, these tactics may be a special protection.

A further, strong argument for the correctness of my assumption lies in the fact that evidently not a single one of the large predators ever attacks in the midst of a dense herd of its prey. Not only do the big predatory animals, such as lions and tigers, hesitate, in the face of the defensive powers of their prey, before leaping onto an African buffalo in the herd, but even smaller hunters of defenseless game try, almost without exception, to separate a single animal from the herd before they attack it. Peregrine and Hobby Falcon have a special movement pattern serving this end alone. W. Beebe observed corresponding behavior in fish in the sea. He saw a big amber jack following a shoal of little porcupine fish, waiting patiently until one of the small fish separated itself from the group to snap up some still smaller prey. Each time this happened, the small fish met its end in the stomach of the big one.

Wandering flocks of starlings make use of the bad marksmanship of the predator to spoil his appetite for catching starlings. If a flock of these birds comes within sight of a flying sparrow hawk or hobby, they press so close together that one can hardly imagine that they can still use their wings. In this formation they fly not away from the predator but after him, finally encircling him, just as an amoeba flows around a particle of nourishment, enclosing it in a little vacuole. Some observers have asserted that by this maneuver the air is sucked from under the wings of the predator, preventing him from flying, still more from attacking. This, of course, is nonsense, but an experience of this kind is certainly unpleasant enough for the predator to act as a deterrent, and this fact lends survival value to the whole behavior mechanism.

Some sociologists are of the opinion that the family is the most primitive form of social aggregation, and that the different forms of communities in the higher animals have arisen from it phylogenetically. This theory may be true of several social insects, such as bees, ants, and termites, and possibly of some mammals too, including the primates to which man belongs, but it cannot be applied generally. The most primitive form of a "society" in the broadest sense of the term is the anonymous flock, of which the shoal of ocean fishes is the most typical example. Inside the shoal, there is no structure of any kind, there is no leader and there are no led, but just a huge collection of like elements. Of course these influence each other mutually, and there are certain very simple forms of "communication" between the individuals of the shoal.

When one of them senses danger and flees, it infects with its mood all the others which have perceived its fear. How far the panic in a big shoal spreads, and whether it is able to make the whole shoal turn and flee, is purely a quantitative question, the answer to which depends on how many individuals become frightened and flee and how intensively they do so.

Stimulus situations which attract the fish can be responded to by a whole shoal, even when only one individual has received the stimuli. The resolute swimming of this individual in one direction draws the other fish with it, and here again it is a question of quantity whether or not the whole shoal is pulled along.

The purely quantitative and, in a sense, democratic action of this process called "social induction" by sociologists means that a school of fish is the less resolute the more individuals it contains and the stronger its herd instinct is. A fish which begins, for any reason, to swim in a certain direction cannot avoid leaving the school and thus finding itself in an isolated position. Here it falls under the influence of all those stimuli calculated to draw it back into the school. The more fish there are swimming in the same direction, as a result of some exogenous stimulus, the more likely they are to draw the school with them; the bigger the school and its consequent counterattraction, the less far its members will swim before they return to the school, drawn as by a magnet. A big school of small and closely herded fish thus presents a lamentable picture of indecision. Again and again a small current of enterprising single fish pushes its way forward like the pseudopodium of an amoeba. The longer such pseudopods become, the thinner they grow, and the stronger becomes their longitudinal tension. Generally the whole advance ends in precipitate flight back to the heart of the school. Watching these indecisive actions, one almost begins to lose faith in democracy and to see the advantage of authoritarian politics.

However, it can be shown by a very simple experiment how little justified this standpoint is. Erich von Holst removed, from a common minnow, the forebrain, which, in this species, is the site of all shoaling reactions. The pithed minnow sees, eats, and swims like a normal fish, its only aberrant behavior property being that it does not mind if it leaves the shoal unaccompanied by other fishes. It lacks the hesitancy of the normal fish, which, even when it very much wants to swim in a certain direction, turns around after its first movements to look at its shoalmates and lets itself be influenced according to whether any others follow it or not. This did not matter to the brainless fish: if it saw food, or had any other reason for doing so, it swam resolutely in a certain direction and—the whole shoal followed it. By virtue of its deficiency, the brainless animal had become the dictator!

The important effect of intra-specific aggression, dispersing and spacing out the animals of a species, is essentially opposed to that of herd attraction. Strong aggression and very close herding exclude one another, but less extreme expressions of the two behavior mechanisms are not incompatible. In many species which form large flocks, the individuals never come nearer to each other than a certain minimum distance; there is always a constant space between every two animals. Starlings, sitting like a string of pearls at exactly regular intervals along a telegraph wire, are a good example of this spacing. The distance between the individuals corresponds exactly to the distance at which two starlings can reach each other with their beaks. Immediately after landing, they sit irregularly distributed, but soon those that are too close together begin to peck at each other and continue to do so until the "prescribed" *individual distance,* as Hediger appropriately called it, is established. We may conceive the space, whose radius is represented by the individual distance, as a very small, movable territory, since the behavior mechanisms ensuring its maintenance are fundamentally the same as those which effect the demarcation of territory. There are also genuine territories, for example in the colony-nesting gannets, arising in the same way as the perching distribution of starlings: the tiny territory of a gannet pair is just big enough to prevent two neighboring birds, in the center of their territories—that is, when they are sitting on their nests—from reaching each other with the tips of their beaks if they stretch out their necks as far as they can.

It is only for the sake of completeness that I have mentioned here that gregariousness and intra-specific aggression are not entirely incompatible. In general, typical herd animals lack any aggressive instinct and with it any individual distance. Herringlike and carplike fish huddle together when disturbed, but also when resting, almost to the point of physical contact; many fish which are territorial and highly aggressive during the reproductive season lose all aggressive behavior when they come together in swarms outside the breeding season. This applies to many cichlids, to sticklebacks, and some others. The nonaggressive psychophysiological state of schooling can usually be deduced from the special color patterns of the fish. Numerous bird species have the habit of retiring, outside the breeding season, into the anonymity of the flock. This is the case with storks, herons, swallows, and a number of songbirds in which there is no bond holding the partners together during the autumn and winter.

Only in a few bird species, for example swans, wild geese, and cranes, do mates or parents and children keep together in the big migrating flocks. The large numbers of birds and the close formation of most big flocks must make it very hard for individual partners not to lose touch. But most species which form large herds, flocks, or schools seem to set no store by such individual contacts. Their form of society is of necessity completely anonymous, every individual is just as content with any one fellow member of the species as with any other, and the bonds of personal friendship which seem so indispensable to us simply do not occur in these species.

The ties which hold such an anonymous flock together are very different indeed from those which lend strength and security to our own society. Nevertheless, one could imagine that personal friendship and love might well have developed in the lap of the peaceful anonymous society, a thought that suggests itself the more readily since the anonymous crowd undoubtedly evolved phylogenetically long before the personal bond. To obviate misunderstanding, I must here anticipate the theme of a later chapter, "The Bond." Anonymous flock formation and personal friendship exclude each other to a large extent because personal friendship is always coupled with aggression. We do not know of a single animal which is capable of personal friendship and which lacks aggression. This combination is particularly impressive in animals which are aggressive only during the reproductive season and which otherwise lack aggression and form anonymous flocks. When such creatures form any personal ties, these are dissolved with the loss of aggression. For this reason, in storks, chaffinches, cichlids, and others, the mates do not remain together when the big anonymous flocks assemble for migration.

To our human mind, personal friendship represents one of the most cherished values, and any social organization not built upon its basis inspires us with a chilling sense of the inhuman. This will become clearer in the next two chapters. However, even the simple and seemingly innocuous mechanisms of anonymous flocking can turn into something not only inhuman but truly terrible. In human society, these mechanisms remain more or less hidden, being superseded by nonanonymous, well-organized relationships between individuals, but there is one contingency in which they erupt with the uncontrollable power of a volcano and gain complete mastery over man, causing behavior that can no longer be called human. This horrible recrudescence of the ancient mechanisms of flocking behavior occurs in mass panic. I was once an unwilling witness of the sudden emergence and rapidly snowballing effect of this process of dehumanization, and if I was not drawn into its vortex it was only because, thanks to my knowledge of flocking behavior, I had seen the approaching danger sooner than others and had had time to guard against my own reactions. To me there is small pride in the memory; on the contrary, no one can put much trust in his own self-mastery who has ever seen men more courageous than himself, men fundamentally disciplined and self-controlled, rushing blindly along, closely huddled, all in the same direction, with eyes protruding, chests heaving, and trampling underfoot everything that comes in their way, exactly like stampeding ungulates, and no more accessible to reason than they.

Chapter Nine

Social Organization without Love

I have contrasted the simple flock organization of the anonymous individuals with the social order built on personal relationships in order to emphasize that these two mechanisms of social behavior do, to a large extent, exclude one another; but this does not in any way imply that there are no other types of social organization. In animals, there are other relationships between certain individuals which bind them for long periods, even for life, without the involvement of personal ties. Just as in human society there are business partners who work happily together but would never think of meeting outside office hours, so there are in many animal species individual ties arising only through a common interest of the partners in a shared "enterprise." The anthropomorphizing animal lover will not be pleased to hear that in many birds, including some that live in lifelong "matrimony," male and female have no interest in each other's company unless they have a common function to fulfill at the nest or in the service of the brood. An extreme case of such a bond in which the partners are bound neither by mutual recognition nor by love is seen in the tie called by Heinroth "local mating." Males and females of the South European Green Lizard, for example, defend their territories against members of the same sex only. The male takes no action against an encroaching female (indeed he cannot because he is prevented from attacking a female by the inhibition described on page 123). The female cannot attack even a young male inferior to herself in strength, because an immense innate respect for the insignia of maleness prevents her from doing so. Thus male and female lizard would demarcate their territories as independently of each other as animals of two entirely different species which keep no intra-specific distance from each other, were it not for the fact that they both show similar "taste" in choosing a dwelling. But neither in our large and well-planned enclosure, forty yards square, nor in nature, is there an unlimited amount of tempting accommodation in the form of hollows between stones, holes in the earth, and other such places. So it is inevitable that sometimes a male and a female, having no individual distance, move into the same dwelling. And since two dwellings are seldom equally useful and attractive, it is not surprising that in our enclosure a certain particularly favorable hollow, facing south, was soon occupied by the strongest male and the strongest female in our lizard colony. Though they do not have a particular preference for each other, animals living in such close contact copulate more frequently with each other than with a chance partner met on the territorial border. If one of the "local mates" was removed experimentally, it soon "got around" that a particularly desirable male, or female, territory was vacant. Then severe territorial fights took place between the interested parties, with the usual result that by next day the next-strongest male or female had taken possession of the dwelling place and of the sexual partner too.

It is astonishing that our white storks should behave almost exactly like the lizards, a fact quite at variance with the gruesome story told wherever white storks nest and sportsmen meet. From time to time, some daily paper reports how the storks, before leaving for Africa, hold a tribunal where all crimes of individual storks are tried, and females guilty of adultery are condemned to death and executed mercilessly. In reality, a stork is not very fond of his wife, and it is very doubtful whether he would even recognize her away from the nest. A stork couple is certainly not held together by that magic elastic band which, in geese, ravens, or jackdaws, evidently pulls harder as the mates get farther away from each other. The male stork and his wife almost never fly together at a fixed distance like the mates of the species mentioned above and many others, and they even migrate at different times. In the spring, the male stork returns to the nesting place much earlier than his wife or, to be more precise, the female who shares the same nest. While he was director of the Rossitten bird observatory, Professor Ernst Schüz made the following enlightening observations of the storks nesting on his roof. One year, the male came back early, and after he had been home for some time and was standing on his nest, a strange female appeared. He greeted her with chattering, and she at once made herself at home on the nest, greeting him in the same way. He admitted her without hesitation and treated her in every detail exactly as a male stork treats his long-awaited wife on her return. Professor Schüz told me he could have sworn that the newcomer was the old female, if the leg bands —or rather, the lack of them—had not taught him better, or perhaps I should say worse.

The two were busily occupied repairing and relining the nest when suddenly the old wife arrived. The two females started an embittered territorial fight which the male watched quite disinterestedly and without attempting to defend his old wife against the new one, or vice versa. Finally the new wife flew off, beaten by the "legitimate" one, and after the change of partners, the male continued with his nesting business exactly where he had been interrupted by the fight of the rivals.

He showed no sign of having noticed the change of wives that had twice taken place. What a remarkable contrast to the myth of the stork tribunal. If such a bird caught his wife *in flagranti,* with the neighbor on the next roof, he probably would not even recognize her as his own!

Night herons behave very similarly to storks, but there are several other species of herons in which, as Otto Koenig has shown, the mates certainly recognize each other individually and associate up to a certain extent even away from the nest. I know night herons well, because for many years I had an artificially settled colony of tame, free-flying birds of this species in my garden and I was thus able to observe their pairing, nest building, brooding, and baby rearing, in minute detail. When two mates met on neutral ground, that is at a certain distance from their common nest territory, whether they were fishing in the pond or coming to be fed in a field about a hundred yards away from their nesting tree, they showed absolutely no signs of recognizing each other. They chased each other just as furiously from a good fishing place, and fought just as angrily over the food I gave them, as any other two utterly unrelated birds. The mates never flew together, and the formation of larger or smaller swarms, when they flew down to the Danube at dusk to fish, bore the character of anonymous flocks.

Equally anonymous is the organization of the nesting colony. It is entirely different from that of the jackdaw colony, which consists of an exclusive circle of old friends. In spring, every night heron in reproductive mood wants to have its nest near, but not too near, that of another bird. One has the impression that some squabbling with a hostile neighbor is essential to the bird in order to get it into a proper nesting mood. Just as in the case of the nest territory of the Gannet, or the sitting place of a starling, the minimum diameter of a nest territory of a night heron is determined by the span of neck and beak of two neighbors, and the centers of two nests can never be nearer than twice the span of a bird's neck and beak. Among the long-necked herons this is a considerable distance.

I cannot say for certain whether the neighbors recognize each other, but I had the impression that a night heron never got used to the approach of another one that had to pass by very closely on the way to its own nest. One would expect that after innumerable repetitions of this procedure, the silly bird would eventually realize that the passer-by, whose nervous look and flattened feathers expressed anything but thirst for conquest, only wanted to squeeze past. But the night heron never understands that his neighbor is himself a territory owner and therefore not dangerous, and he cannot differentiate between the neighbor and a stranger who is a potential usurper of territory. Even the nonanthropomorphizing observer cannot help being irritated by the ever recurring clamor and spiteful beak duels that take place, day and night, in a night heron colony. One would think that this unnecessary waste of energy could easily be obviated, for night herons are fundamentally able to recognize fellow members of their species. The young of a brood know each other even as tiny nestlings and attack any strange night heron baby, even one of the same age, if it is introduced into the nest. The fledglings keep together for a long time, seek protection from each other, and, standing back to back, defend themselves against attack. It is thus the more astonishing that nesting night herons never treat the owners of bordering territories "as though they knew" that these were settled householders with no intention of robbing a neighbor of his territory.

Why, one asks, has the night heron never "hit on the idea" of making use of his proved ability to recognize his fellows for the purpose of selective habituation to his territorial neighbors, thus saving himself an incredible amount of energy and annoyance? The question is difficult to answer, and probably it is asked wrongly. In nature we find not only that which is expedient, but also everything which is not *so* inexpedient as to endanger the existence of the species.

Surprisingly, there is a fish which is capable of doing what the night heron cannot do: it can accustom itself to its settled and harmless neighbor and so avoid the elicitation of unnecessary aggression. This fish is a member of a group well known for breaking fish records: the cichlids. In the North African oasis, Gafsah, there lives a little Mouthbreeding Cichlid about whose social behavior we have learned through the close field observations of Rosl Kirchshofer. The males build a closely knit colony of nests, or rather, spawning hollows, in which the females lay their eggs. As soon as they are all fertilized, the female takes them in her mouth and transports them to other places, hatching them in shallow water, thickly grown with plants, where eventually the young are reared. Every male possesses only a relatively tiny territory, and this is almost completely filled by the spawning hollow which the fish constructs by fanning with his tail and digging with his mouth. Every male tries to entice every passing female into his hollow, by certain ritualized courtship movements and by guiding the female to his own nest. The males spend a very large part of the year performing these tactics, in fact it is possible that they are in the spawning place all the year round. There is no reason to suppose that they often change their territory, every fish has plenty of time to get to know his neighbor, and it has long been known that cichlids are capable of this. Dr. Kirchshofer performed the laborious task of catching all the males of a nesting colony and marking them individually. Every fish knew the owners of neighboring territories very exactly and tolerated them peacefully at closest quarters, while he immediately attacked every stranger which approached his spawning hollow even from farther away.

This peacefulness of the male mouthbreeders of Gafsah, depending on individual recognition of their fellows, is not yet that bond of friendship which we shall describe in the next chapter but one. In these fish, there is not yet that attraction between personally acquainted individuals which keeps them permanently together—and this is the objectively demonstrable sign of friendship. But in a field of forces, in which mutual repulsion is ever present, every lessening of the active repulsive force between two particular objects has consequences which are, in effect, equivalent to attraction. In still another respect, the nonaggression pact of neighboring mouthbreeder males resembles true friendship: the lessening of repulsion, as well as the attraction of being friends, depends on the degree of acquaintanceship of the individuals concerned. Selective habituation to all stimuli emanating from individually known members of the species is probably the prerequisite for the origin of every personal bond, and it is probably its precursor in the phylogenetic evolution of social behavior.

Generally, other conditions being equal, mere acquaintanceship with a fellow member of the species exerts a remarkably strong inhibitory effect on aggressive behavior. In human beings, this phenomenon can regularly be observed in railway carriages, incidentally an excellent place in which to study the function of aggression in the spacing out of territories. All the rude behavior patterns serving for the repulsion of seat competitors and intruders, such as covering empty places with coats or bags, putting up one's feet, or pretending to be asleep, are brought into action against the unknown individual only. As soon as the newcomer turns out to be even the merest acquaintance, they disappear and are replaced by rather shamefaced politeness.

Chapter Ten

Rats

There is a type of social organization characterized by a form of aggression that we have not yet encountered: the collective aggression of one community against another. I will try to show how the misfunctioning of this social form of intra-specific aggression constitutes "evil" in the real sense of the word, and how the kind of social order now to be discussed represents a model in which we can see some of the dangers threatening ourselves.

In their behavior toward members of their own community, the animals here to be described are models of social virtue; but they change into horrible brutes as soon as they encounter members of any other society of their own species. Communities of this type possess too many individual animals for these to know each other personally, and in most cases membership of a certain society is identifiable by a definite smell, common to all members.

In the huge communities of social insects it has long been known that their societies, often comprising millions, are basically families consisting of the descendants of a single female or pair which founded the colony. It is also well known that among bees, termites, and ants the members of such a large clan recognize each other by a characteristic hive, nest, or anthill smell, and that murder occurs if a member of a strange colony inadvertently enters the nest. Massacres ensue if a human experimenter inhumanly tries to mix two colonies. It has, I think, been known only since 1950 that there are large families of rodents which behave similarly. F. Steiniger and J. Eibl-Eibesfeldt made this important discovery at about the same time but independently of each other, Steiniger in the Brown Rat and Eibl-Eibesfeldt in the House Mouse.

Eibl, who at that time was with Otto Koenig at the Wilhelminenberg Biological Station, worked on the sound principle of living in as close contact as possible with his experimental animals; the house mice, which lived free in his hut, were regularly fed by him, and he moved about quietly and carefully so that they were soon tame enough for him to make observations at close quarters. One day he opened a large container in which he had bred a number of big, wild-colored laboratory mice, not too different from the wild form. As soon as these mice dared to leave the cage and run about in the room, they were attacked furiously by the resident wild mice, and only after hard fighting did they manage to regain the safety of their prison, which they defended successfully against invasion by the wild mice.

Steiniger put brown rats from different localities into a large enclosure which provided them with completely natural living conditions. At first the individual animals seemed afraid of each other; they were not in an aggressive mood, but they bit each other if they met by chance, particularly if two were driven toward each other along one side of the enclosure, so that they collided at speed. However, they became really aggressive only when they began to settle and take possession of territories. At the same time, pair formation started between unacquainted rats from different localities. If several pairs were formed at the same time the ensuing fights might last a long time, but if one pair was formed before the others had started, the tyranny of the united forces of the two partners increased the pressure on the unfortunate co-tenants of the enclosure so much that any further pair formation was prevented. The unpaired rats sank noticeably in rank and were constantly pursued by the two mates. Even in the 102-square-yard enclosure, two or three weeks sufficed for such a pair to kill all the other residents, ten to fifteen strong adult rats.

The male and female of the victorious pair were equally cruel to their subordinates, but it was plain that he preferred biting males and she females. The subjugated rats scarcely defended themselves, made desperate attempts to flee, and in their desperation took a direction which rarely brings safety to rats, namely upwards. Steiniger repeatedly saw weary, wounded rats sitting exposed and in broad daylight high up in bushes and trees, evidently outside occupied territory. The wounds were usually on the end of the back and on the tail, where the pursuer had seized them. Death was seldom caused by sudden, deep wounds or loss of blood but more frequently by sepsis, particularly in the case of bites which penetrated the peritoneum. But usually the animals died of exhaustion and nervous overstimulation leading to disturbance of the adrenal glands.

A particularly effective and cunning method of killing fellow members of the species was observed by Steiniger in female rats, which became veritable murder specialists. He writes, "They slink up stealthily, suddenly spring at their victim, which is perhaps eating innocently at the feeding place, and bite it in the side of the neck, frequently injuring the carotid artery. The fight usually lasts only a few seconds, the mortally wounded animal bleeds internally, and on postmortem profuse hemorrhages are found subcutaneously and in the body cavities."

Having witnessed the bloody tragedies which enabled the

surviving couple finally to rule the whole enclosure, one would hardly expect to see the development of the society which is soon built up by the victorious murderers. The tolerance, the tenderness which characterizes the relation of mammal mothers to their children, extends in the case of the rats not only to the fathers but to all grandparents, uncles, aunts, cousins, and so on. The mothers put their children into the same nest, and it is improbable that each mother tends only her own offspring. There are no serious fights within the large family even when this comprises dozens of animals. In the wolf pack, whose members are otherwise so considerate of each other, the highest-ranking animals eat first from the common prey. But in the rat pack there is no ranking order, the pack attacks its prey in a body, and the strongest animals play the chief part in overcoming it. But in eating it, according to Steiniger, "The smaller animals are the most forward, the larger ones good-humoredly allow the smaller ones to take pieces of food away from them. In reproduction, too, the more lively half and three-quarter grown animals usually take precedence of the adults. All rights are open to them, and even the strongest adult puts nothing in their way."

Within the pack there is no real fighting, at the most there is slight friction, boxing with the fore-paws or kicking with the hind paws, but never biting; and within the pack there is no individual distance, on the contrary, rats are contact animals in the sense of Hediger, and they like touching each other. The ceremony of friendly contact is the so-called "creeping under," which is performed particularly by young animals while larger animals show their sympathy for smaller ones by creeping over them. It is interesting that overdemonstrativeness in this respect is the most frequent cause of harmless quarrels within the big family. When an older animal which happens to be eating is importuned too much by a younger one, it repels it by boxing and kicking.

Within the pack there is a quick news system functioning by mood transmission, and, what is most important, there is a conservation and traditional passing on of acquired experience. If the rats find a hitherto unknown food, according to Steiniger the first rat to find it usually decides whether or not the family should eat it. "If a few animals of the pack pass the food without eating any, no other pack member will eat any either. If the first rats do not eat poisoned bait, they sprinkle it with their urine or feces. Even when, owing to local conditions, it is extremely uncomfortable to deposit feces on top of the poison, nevertheless it is often done." But the most astonishing fact is that knowledge of the danger of a certain bait is transmitted from generation to generation and the knowledge long outlives those individuals which first made the experience. The difficulty of effectively combating the most successful biological opponent to man, the Brown Rat, lies chiefly in the fact that the rat operates basically with the same methods as those of man, by traditional transmission of experience and its dissemination within the close community.

Serious fights between members of the same big family occur in one situation only, which in many respects is significant and interesting: such fights take place when a strange rat is present and has aroused intra-specific, inter-family aggression. What rats do when a member of a strange rat clan enters their territory or is put in there by a human experimenter is one of the most horrible and repulsive things which can be observed in animals. The strange rat may run around for minutes on end without having any idea of the terrible fate awaiting it; and the resident rats may continue for an equally long time with their ordinary affairs till finally the stranger comes close enough to one of them for it to get wind of the intruder. The information is transmitted like an electric shock through the resident rat, and at once the whole colony is alarmed by a process of mood transmission which is communicated in the

Brown Rat by expression movements but in the House Rat by a sharp, shrill, satanic cry which is taken up by all members of the tribe within earshot.

With their eyes bulging from their sockets, their hair standing on end, the rats set out on the rat hunt. They are so angry that if two of them meet they bite each other. "So they fight for three to five seconds," reports Steiniger, "then with necks outstretched they sniff each other thoroughly and afterwards part peacefully. On the day of persecution of the strange rat all the members of the clan are irritable and suspicious." Evidently the members of a rat clan do not know each other personally, as jackdaws, geese, and monkeys do, but they recognize each other by the clan smell, like bees and other insects. A member of the clan can be branded as a hated stranger, or vice versa, if its smell has been influenced one way or the other. Eibl removed a rat from a colony and put it in another terrarium specially prepared for the purpose. On its return to the clan enclosure a few days later, it was treated as a stranger, but if the rat was put, together with some soil, nest, etc., from this clan enclosure, into a clean, empty battery jar so that it took with it a dowry of objects impregnated with a clan smell, it would be recognized afterwards, even after an absence of weeks.

Heart-breaking was the fate of a house rat which Eibl had treated in the first way, and which in my presence he put back into the clan enclosure. This animal had obviously not forgotten the smell of the clan, but it did not know that its own smell was changed. So it felt perfectly safe and at home, and the cruel bites of its former friends came as a complete surprise to it. Even after several nasty wounds, it did not react with fear and desperate flight attempts, as really strange rats do at the first meeting with an aggressive member of the resident clan. To softhearted readers I give the assurance, to biologists I admit hesitatingly, that in this case we did not await the bitter end but put the experimental animal into a protective cage which we then placed in the clan enclosure for repatriation.

Without such sentimental interference, the fate of the strange rat would be sealed. The best thing that can happen to it is, as S. A. Barnett has observed in individual cases, that it should die of shock. Otherwise it is slowly torn to pieces by its fellows. Only rarely does one see an animal in such desperation and panic, so conscious of the inevitability of a terrible death, as a rat which is about to be slain by rats. It ceases to defend itself. One cannot help comparing this behavior with what happens when a rat faces a large predator that has driven it into a corner whence there is no more escape than from the rats of a strange clan. In the face of death, it meets the eating enemy with attack, the best method of defense, and springs at it with the shrill war-cry of its species.

What is the purpose of group hate between rat clans? What species-preserving function has caused its evolution? The disturbing thought for the human race is that this good old Darwinian train of thought can only be applied where the causes which induce selection derive from the extra-specific environment. Only then does selection bring about adaptation. But wherever competition between members of a species effects sexual selection, there is, as we already know, grave danger that members of a species may in demented competition drive each other into the most stupid blind alley of evolution. On page 41 we have read of the wings of the Argus pheasant and the working pace of Western civilized man as examples of such errors of evolution. It is thus quite possible that the group hate between rat clans is really a diabolical invention which serves no purpose. On the other hand it is not impossible that as yet unknown external selection factors are still at work; we can, however, maintain with certainty that those indispensable species-preserving functions of intra-specific aggression, which have been discussed in Chapter Four, are not served by clan fights. These serve neither spatial distribution nor the selection of strong family defenders—for among rats these are seldom the fathers of the descendants—nor any of the functions enumerated in Chapter Three.

It can readily be seen that the constant warfare between large neighboring families of rats must exert a huge selection pressure in the direction of an ever increasing ability to fight, and that a rat clan which cannot keep up in this respect must soon fall victim to extermination. Probably natural selection has put a premium on the most highly populated families, since the members of a clan evidently assist each other in fights against strangers, and thus a smaller clan is at a disadvantage in fights against a larger one. On the small North Sea island of Norderoog, Steiniger found that the ground was divided between a small number of rat clans separated by a strip of about fifty yards of no rat's land where fights were constantly taking place. The front is relatively larger for a small clan than for a big one, and the small one is therefore at a disadvantage.

Chapter Eleven

The Bond

In the three different types of social order described in the foregoing chapters, relations between individual beings are completely impersonal. It is characteristic of the supra-individual community that one individual can be exchanged for almost any other. We have seen the first trace of personal relationships in the territory-owning males of the Gafsah mouthbreeders, which form nonaggression pacts with their neighbors and are aggressive only toward strange intruders. However, this is only a passive tolerance of the known neighbor. Neither of the individuals yet has for the other an attraction that could cause him to follow if the partner should swim away, or, should the partner stay in one place, to stay there too for his sake, or still less to search for him actively should he disappear. These behavior patterns of an objectively demonstrable mutual attachment constitute the personal tie which is the subject of this chapter. From now on, I will call it the bond, and the society which it holds together, the group. The group is thus characterized by the fact that, like the anonymous crowd, it is held together by reactions elicited by one member in another, but in contrast to the impersonal social order, the attachment reactions are inseparably linked with the individualities of group members.

As in the mutual-tolerance pact of Gafsah mouthbreeders, it is a prerequisite of group formation that individual animals should be capable of reacting selectively to the individuality of every other member; but those mouthbreeders which react in a different way to the neighbor and to the stranger do so in one place only, in their own nest hollow, and a number of additional circumstances are involved in this process of special habituation. It is open to question whether the fish would treat the neighbor in the same way if they both suddenly found themselves in an unfamiliar place; but true group formation is characterized by its independence of place. The part which every member plays in the life of every other one remains the same in an amazing number of different environmental situations, that is to say, personal recognition of the partner in all possible conditions of life is the essential for every group formation. Recognition of the partner must always be learned individually.

When we consider the series of life patterns in the ascending scale, from the simpler to the higher, we encounter group formation in the above sense for the first time among higher teleosteans, namely among Spiny Rayed Fish, particularly Cichlids, in the closely related fish of the perch family, the Demoiselles, Angelfish, and Butterfly Fish. We have met with these three families of tropical marine fish in the first two chapters, and it is significant that they possess a particularly large measure of intra-specific aggression.

In discussing "anonymous flock" formation, I have stressed that this most widespread and most primitive form of social order did not arise from the family, the unit of parents and children, as it did in the case of the quarrelsome rat clans and in the packs of other mammals. In a rather different sense, the phylogenetic prototype of the personal bond and of group formation is the attachment between two partners which together tend their young. From such a tie a family can easily arise, but the bond with which we are here concerned is of a much more special kind. We will now describe how this bond comes about in Cichlids.

In observing, with a thorough knowledge of all the expression movements, the processes which in Cichlids effect the coming together of partners of opposite sex, it is a nerve-racking experience to see the prospective mates in a state of real fury with each other. Again and again they are close to starting a vicious fight, again and again the ominous flare-up of the aggressive drive is only just inhibited and murder sidestepped by a hair. Our apprehension is by no means founded on a false interpretation of the particular expression movements observed in our fish: every fish breeder knows that it is risky to put male and female of a Cichlid species together in a tank, and that there is considerable danger of casualty if pair formation is not constantly supervised.

Under natural conditions, habituation is largely responsible for preventing hostilities between the prospective mates. We can best imitate natural conditions by putting several young, still peaceable fishes in a large aquarium and letting them grow up together. Pair formation then takes place in the following way: on reaching sexual maturity a certain fish, usually a male, takes possession of a territory and drives out all the others. Later, when a female is willing to pair, she approaches the territory owner cautiously and, if she acknowledges the superior rank of the male, responds to his attacks which, at first, are quite seriously meant, in the way described on page 104, with the so-called "coyness behavior," consisting, as we already know, of behavior elements arising partly from mating and partly from escape drives. If, despite the clearly aggression-inhibiting intention of these gestures, the male attacks, the female may leave his territory for a short time, but sooner or later she returns. This is repeated over a varying period until each of the two animals is so accustomed to the presence of the other that the aggression-eliciting stimuli inevitably proceeding from the female lose their effect.

As in many similar processes of specific habituation, here, too, all fortuitously occurring accessory factors become part of the entire situation to which the animal finally becomes accustomed. If any of these factors is missing, the whole effect of the habituation will be upset. This applies in particular to the beginning stages of peaceful cohabitation, when the partner must always appear on the accustomed route, from the accustomed side; the lighting must always be the same, and so on, otherwise each fish considers the other as a fight-releasing stranger. Transference to another aquarium can at this stage completely upset pair formation. The closer the acquaintanceship, the more the picture of the partner becomes independent of its background, a process well known to the Gestalt psychologist as also to the investigators of conditioned reflexes. Finally, the bond with the partner becomes so independent of accidental conditions that pairs can be transferred, even transported far away, without rupture of their bond. At most, pair formation "regresses" under these circumstances, that is, ceremonies of courtship and appeasement may recur, which in long-mated partners had long ago disappeared, having ceded to force of habit.

If pair formation runs an undisturbed course, the male's sexual behavior gradually comes to the fore. There may already be traces of these behavior patterns in his first seriously intended attacks on the female, but now they increase in intensity and frequency without, however, causing the disappearance of the expression movements implying aggressive mood. In the female, however, the original escape-readiness and "submissiveness" decrease very quickly. Movements expressive of fear or escape mood disappear in the female more and more with the consolidation of pair formation, in fact they sometimes disappear so quickly that, during my early observations of Cichlids, I overlooked them altogether and for years erroneously believed that no ranking order existed between the partners of this family. We have already heard (page 103) what part ranking order plays in the mutual recognition of the sexes, and it persists latently, even when the female has completely stopped making submissive gestures to her mate. Only on the rare occasions when an old pair quarrels does she do it again.

At first nervously submissive, the female gradually loses her fear of the male, and with it every inhibition against showing aggressive behavior, so that one day her initial shyness is gone and she stands, fearless and truculent, in the middle of the territory of her mate, her fins outspread in an attitude of self-display, and wearing a dress which, in some species, is scarcely distinguishable from that of the male. As may be expected, the male gets furious, for the stimulus situation presented by the female lacks nothing of the key stimuli which, from experimental stimulus analysis, we know to be strongly fight-releasing. So he also assumes an attitude of broadside display, discharges some tail beats, then rushes at his mate, and for fractions of a second it looks as if he will ram her—and then the thing happens which prompted me to write this book: the male does not waste time replying to the threatening of the female; he is far too excited for that, he actually launches a furious attack which, however, *is not directed at his mate but, passing her by narrowly, finds its goal in another member of his species*. Under natural conditions this is regularly the territorial neighbor.

This is a classical example of the process which we call, with Tinbergen, a *redirected activity*. It is characterized by the fact that an activity is released by *one* object but discharged at *another*, because the first one, while presenting stimuli specifically eliciting the response, simultaneously emits others which inhibit its discharge. A human example is furnished by the man who is very angry with someone and hits the table instead of the other man's jaw because inhibition prevents him from doing so, although his pent-up anger, like the pressure within a volcano, demands outlet. Most of the known cases of redirected activity concern aggressive behavior elicited by an object which simultaneously evokes fear. In this special case, which he called "bicycling," B. Grzimek first recognized and described the principle of redirection. The "bicyclist" in this case is the man who bows to his superior and treads on his inferior. The mechanism effecting this behavior is particularly clear when an animal approaches its opponent from some distance, then, on drawing near, notices how terrifying the latter really is, and now, since it cannot check the already started attack, vents its anger on some innocent bystander or even on some inanimate substitute object.

There are, of course, innumerable further forms of redirected movements, and various combinations of opposing drives can produce them. The special case of the Cichlid male is very significant for our theme because analogous processes play a decisive role in the family and social life of a great many higher animals and man. The problem of how to prevent inter-marital fighting is solved in a truly remarkable way not only by not inhibiting the aggression elicited in each partner by the presence of the other, but by putting it to use in fighting the hostile neighbor. This solution has evidently been found independently in several unrelated groups of vertebrates.

The averting of the undesirable aggression elicited by the partner and its canalizing in the desired direction of the territorial neighbor is, in the dramatic case of the cichlid male, no momentary decision which the animal can make or not make at the critical moment. It has long been ritualized and has become a part of the fixed instinct inventory of the particular species. Everything that we have learned, in Chapter Five, about the process of ritualization helps us to understand that from the redirected activity a new instinct movement can arise which, like all others, must find its discharge, and hence presents a need, an independent motive for action.

In prehistoric times, probably around the Chalk Age (a million years or so make no difference here), a similar case of redirected activity must have happened by chance, just as the tobacco smoking of the Red Indian chiefs described in Chapter Five happened by chance, otherwise no rite could have arisen. One of the great constructors of the change in species, selection, always requires some fortuitously arising material to work on, and its blind but busy colleague, mutation, provides the material.

As with many physical characteristics and many instinctive motor patterns, the individual development, the ontogeny, of a ritualized ceremony follows roughly the path taken by its phylogeny. To be exact, it repeats the ontogeny which the same character took in the ancestral forms, as Carl Ernst von Baer rightly recognized. However, for our purpose the wider definition suffices. The rites evolved from the redirected attack resemble its unritualized prototype far more in its first appearance than in its later form. In a newly paired cichlid male, when the intensity of the whole reaction is not too pronounced, it can clearly be seen that the male would like to ram his young mate but that at the last moment he is prevented by other motives from doing so, and that he now prefers to vent his anger on his neighbor. In the fully developed ceremony, the "symbol" has become further removed from the symbolized, and its origin is veiled by the "theatrical" effect of the whole reaction, as also by the fact that it is obviously performed for its own sake. Thus its function and symbolism become much more apparent than its origin. A more exact analysis is necessary in order to find out how many of the originally conflicting drives are still present in the individual case. A quarter of a century ago, when my friend Alfred Seitz and I first became acquainted with this rite, we soon understood the function of the "nest-relief" and "greeting" ceremonies of cichlids, but for a long time we did not understand its phylogenetic origin.

However, in the first exactly examined species, the African Jewel Fish, we were immediately struck by the resemblance between the gestures of threatening and of "greeting." We soon learned to differentiate between them and to predict correctly whether the particular movements would lead to fighting or pair formation, but to our dismay we were unable to find out which were the salient points for our verdict. Only when we made a closer analysis of the precarious transitions between serious threatening of the mate and the appeasement ceremony did the difference become clear. In threatening, the fish stops suddenly exactly beside the threatened opponent, particularly when it is excited enough to perform not only broadside display but also the sideways tail beat. Conversely, in the appeasing ceremony he not only does not stop opposite the partner but he swims past her, emphatically exaggerating his forward movements, at the same time directing his broadside display and tail beat toward her. The direction in which the fish presents its ceremony is strikingly different from that in which it sets itself in motion for attack. If before the ceremony it has been standing still in the water near the mate, it always begins to swim forward resolutely before performing the broadside display and tail beat. Thus it is very clearly "symbolized" that the mate is not the object of his attack but that his goal is to be found somewhere else, further away in the direction in which the fish is swimming.

The so-called *functional change* is a means often used by the two great constructors of evolution to put to new purposes remnants of an organization whose function has been outstripped by the progress of evolution. With daring fantasy, the constructors have, for example, made from a water-conducting gill slit an air-containing, sound-conducting hearing tube; from two bones of the jaw joint two little auditory bones; from a parietal eye an endocrine gland, the pineal body; from a reptile's arm a bird's wing, and so on. However, all these amazing metamorphoses seem tame in comparison with the ingenious feat of transforming, by the comparatively simple means of redirection and ritualization, a behavior pattern which not only in its prototype but even in its present form is partly motivated by aggression, into a means of appeasement and further into a love ceremony which forms a strong tie between those that participate in it. This means neither more nor less than converting the mutually repelling effect of aggression into its opposite. Like the performance of any other independent instinctive act, that of the ritual has become a need for the animal, in other words an end in itself. Unlike the autonomous instinct of aggression, out of which it arose, it cannot be indiscriminately discharged at any anonymous fellow member of the species, but demands for its object the personally known partner. Thus it forms a *bond* between individuals.

We must consider what an apparently insoluble problem is here solved in the simplest, most elegant and complete manner: two furiously aggressive animals, which in their appearance, coloring, and behavior are to each other what the red rag (though only proverbially) is to the bull, must be made to agree within the narrowest space, at the nesting place, that is at the very place which each regards as the center of its territory, where intra-specific aggression is at its peak. And this in itself difficult task is made more difficult by the additional demand that intra-specific aggression must not be weakened in either of the partners. We know from Chapter Three that even the slightest decrease in aggression toward the neighboring fellow member of the species must be paid for with loss of territory and, at the same time, of sources of food for the expected progeny. Under these circumstances, the species "cannot afford," for the sake of preventing mate fights, to resort to appeasement ceremonies such as submissive or infantile gestures whose prerequisite is reduction of aggression. Ritualized redirection precludes not only this undesirable effect, but moreover makes use of the key stimuli proceeding from one mate to stimulate the other against the territorial neighbor. I consider this behavior mechanism supremely ingenious, and much more chivalrous than the reverse analogous behavior of the man who, angry with his employer during the day, discharges his pent-up irritation on his unfortunate wife in the evening.

In the great family tree of life, a particularly successful solution is often found by the different branches, independently of one another. Insects, fishes, birds, and bats have "invented" wings; squids, fishes, ichthyosaurs, and whales, the torpedo form. So it is not surprising that fight-preventing behavior mechanisms based on ritualized redirection of attack occur in analogous developments in many different animals.

There is, for example, the marvelous appeasement ceremony, generally known as the "dance" of cranes, which, when the symbolism of its behavior patterns is fully understood, tempts us to translate it into human language. A bird rears up before another one, unfolding its mighty wings, its beak pointing toward the other bird, its eyes fixed piercingly on him, the very image of ominous threatening; so far the appeasement gesture resembles the preparation for attack; but the next moment the bird turns this exhibition of his own fearfulness away from his opponent by a right about turn; and now, still with widely spread wings, he presents to his partner his defenseless occiput, which, in the European crane and many other species, is decorated with a little ruby-red cap. For seconds, the "dancing" bird remains in this position, expressing in easily understood symbolism that his threat of attack is emphatically not directed against his partner but, on the contrary, away from him, against the wicked world outside, implying in this manner the motive of comradely defense. Now the crane turns again toward his friend and repeats this demonstration of his size and strength, only quickly to turn around once more and perform emphatically a fake attack on any substitute object, preferably a nearby crane which is not a friend, or even on a harmless goose or on a piece of wood or stone which he seizes with his beak and throws three or four times into the air. The whole procedure says as clearly as human words, "I am big and threatening, but not toward you—toward the other, the other, the other."

Less dramatic in its sign language but much more significant is the appeasement ceremony of Anatidae, ducks, geese, and swans, called by Oskar Heinroth the *triumph ceremony*. The special significance of this rite for our theme lies in the fact that it is seen in different stages of development and complexity in various representatives of the above-named bird group. From this gradation we can form a picture of how, in the course of phylogeny, an anger-diverting gesture of embarrassment has developed into a bond which shows a mysterious relationship to that other bond between human beings which seems to us the strongest and most beautiful on earth.

In its most primitive form, seen in the so-called rab-rab palaver of the Mallard, threatening differs very little from greeting. The shade of difference in the orientation of the rab-rab chatter when it is a matter of threatening on the one hand or of greeting on the other became clear to me only when I had learned to understand the principle of the redirected appeasement ceremony, through the more exact study of cichlids and geese, in which species it is easier to understand. Ducks face each other with beaks raised just above the horizontal and utter very quickly the bisyllabic call note, rendered in the drake as "rab-rab," in the duck as the more nasal "quang-wang, quangwang." As we already know, it is not only social inhibiting mechanisms which can cause a deviation from the threatening direction; fear of the object can have the same effect. For this reason, two threatening drakes standing opposite each other and not quite daring to attack do not point their bills directly at each other while performing, with raised chins, their rab-rab chatter. In the event of their really pointing them, it may be predicted that they will come to blows immediately, seizing each other by the breast feathers and striking with the shoulder of the wing.

When, however, the drake performs the rab-rab palaver with his mate, particularly when he does so in answer to her inciting movements (pages 64 ff), the turning away of his head is much more marked, its angle increasing with the intensity of the whole rite. In extreme cases this may impel him, still palavering, to turn the back of his head to the female, a gesture corresponding formally to the appeasement ceremony of sea gulls; in these, however, the ceremony has undoubtedly arisen in the manner described on page 134, and definitely not from a redirected activity. This fact should act as a warning against too ready homologizing! From this head turning of the drake further ritualization has evolved the gesture of presenting the back of the head, peculiar to many duck species, which plays a big part in courtship in Mallards, Teals, Pintails, and other Dabbling Ducks, also in the Eiderduck group. The partners of a mallard pair celebrate the ceremony of rab-rab palaver with greatest intensity when they find each other again after a prolonged separation. The same applies to the appeasing gesture with broadside display and tail beat as we know it in cichlid pairs. Because this ceremony so often occurs at the reunion of previously separated partners, early observers interpreted it as "greeting."

Although this interpretation may be true of certain specialized ceremonies of this kind, the frequency and intensity of these gestures at the reunion of partners signifies that originally they had another meaning: the blunting of all aggressive reactions, brought about by habituation to the partner, is partly nullified by even a short interruption of the habituated stimulus situation. Highly impressive examples of this phenomenon can be seen if we let a crowd of aggressive organisms, such as cichlids, Siamese fighting fish, or shama thrushes, grow up amicably together, thus ensuring a high degree of mutual habituation preventing the outbreak of hostilities, and if we then remove an individual for a short time, even for an hour, and afterward return it to the others: on the slightest provocation aggressive behavior surges up like water in delayed boiling.

Other very small changes in the over-all situation can suddenly invalidate the habituation. In the summer of 1961 my old pair of shama thrushes still tolerated a son from their first brood, which inhabited a cage in the same room as their nesting box till long after the time when these birds normally chase their grown-up progeny out of the territory. But when I transferred his cage from the table to a bookcase, the parents began to attack him so intensively that they forgot to fly outside and get food for the babies that they were rearing at the time. Such sudden disintegration of the fighting inhibitions dependent on habituation is apparently a danger that threatens the bonds of partners every time they are separated even for a short time; obviously the strongly pronounced appeasement ceremony, seen at every reunion, is performed for no other reason than to preclude this danger. The fact that "greeting" is the more excited and intense the longer the separation also agrees with this supposition.

Probably our human laughter in its original form was also an appeasement or greeting ceremony. Smiling and laughing in my opinion represent different intensities of the same behavior pattern, that is, they respond with different thresholds to the same particular quality of excitation. In our nearest relations, the chimpanzee and the gorilla, there is unfortunately no greeting movement corresponding in form and function to laughter, but it is seen in many macaques which, as an appeasement gesture, bare their teeth and at intervals turn their heads to and fro, smacking their lips and laying back their ears. It is remarkable that many Orientals smile in the same way when greeting but the most interesting fact is that, while smiling most intensely, they turn their heads a little sideways so that the eyes do not look straight at the person being politely greeted, but past him. In a purely functional consideration of this ritual, it is unimportant how much of its form is fixed by heredity and how much by the cultural tradition of politeness.

In any case, it is tempting to interpret the greeting smile as an appeasing ceremony which, analogously to the triumph ceremony of geese, has evolved by ritualization of redirected threatening. The friendly tooth baring of very polite Japanese lends support to this theory. It is also supported by the fact that in genuinely emotional, intensive greeting between two friends, the smile surprisingly becomes a loud laugh. On considering one's own feelings it seems incongruous that, when meeting a friend after a long separation, the roar of laughter breaks forth unexpectedly from the depths of instinctive strata of our personality. This behavior of two reunited human beings must inevitably remind an objective behavior investigator of the triumph ceremony of greylag geese.

In many respects, the eliciting situations are analogous. When several fairly primitive individuals, such as small boys, laugh together at one or several others not belonging to the same group, the activity, like that of other redirected appeasement ceremonies, contains quite a large measure of aggression directed toward nonmembers of the group. Most jokes provoke laughter by building up a tension which is then suddenly and unexpectedly exploded. Something very similar may happen in the greeting ceremonies of many animals: dogs and geese, and probably other animals, break into intensive greeting when an unpleasantly tense conflict situation is suddenly relieved. The third analogy lies in the fact that laughter, like greeting, tends to create a bond. From self-observation I can safely assert that shared laughter not only diverts aggression but also produces a feeling of social unity.

The simple prevention of fighting may be the original and, in many cases, the chief function of all the rites described above; moreover, even at the relatively low stage of development seen in the rab-rab palaver of the mallard, rites already have so much autonomy that they are aspired to for their own sake. When a mallard drake, constantly uttering his long-drawn monosyllabic call note, "raaaab, raaaab," seeks his mate and, having found her, works himself into a frenzy of rab-rab palaver, raising his chin and presenting the back of his head, the observer cannot refrain from the subjective interpretation that he is delighted to have found her and that his diligent seeking was largely motivated by his longing to indulge in the greeting ceremony. In the more highly ritualized forms of the real triumph ceremony, such as are found in sheldrake and particularly in geese, this impression is even stronger and we are tempted to omit the quotation marks in the use of the word "greeting."

In virtually all dabbling ducks, also in the Common Sheldrake, which in respect of the rab-rab palaver is closely akin to dabbling ducks, this rite has a second function, in the exercise of which only the male performs the appeasement ceremony while the female incites him. A subtle motivation analysis tells us that in this case the male, while directing his threatening gestures toward a neighboring male of the same species, is, to a certain extent, also aggressive toward his own female, whereas she feels no aggression toward him, but genuinely does so toward the stranger. This rite, a combination of redirected threatening of the male and inciting of the female, is functionally analogous to the triumph ceremony in which both partners threaten past each other. In the European Widgeon and the Common Sheldrake, it has developed independently into a particularly beautiful rite. The Chiloe Widgeon, on the other hand, has an equally highly differentiated ceremony, much more closely analogous to the triumph ceremony, since both mates threaten in redirection just as do true geese and most larger forms of Sheldrake. The female of the Chiloe Widgeon wears the male plumage with its glossy green head and light red-brown breast, a unique case among dabbling ducks.

In the Ruddy Sheldrake, the Egyptian Goose, and many related species, the female has a homologous inciting movement, but the male reacts to it less with a ritualized threatening past his wife than with a real, active attack on the neighbor marked by her as hostile. Once this enemy is overpowered, or when the fight has ended at least in a draw, an elaborate triumph ceremony follows. In many species, for example in the Orinoco Goose, the Andean Goose, and others, this ceremony not only produces some remarkable sound effects owing to the difference of the male and female voices, but it is also a very amusing spectacle thanks to the exaggerated miming of the gestures. My film of an Andean Goose couple winning a decisive victory over my friend Niko Tinbergen is guaranteed to bring the house down. First the female, by a short feigned attack, urges her mate against the famous ethologist; and now, gradually working himself up, the gander really attacks, thereby getting into such a fury and beating so angrily with his wings that it looks really convincing when he finally puts Niko to flight. Afterward, Niko's legs and forearms, with which he had warded off the gander, were black and blue from blows and pinches. After the disappearance of the human enemy, a virtually endless triumph ceremony follows which is extraordinarily funny in the exuberance of its all too human expression.

The Egyptian Goose is more energetic than any other female of the Sheldrake group in inciting her mate against all members of the species within reach and, in the absence of these, against birds of other species, much to the dismay of the keeper who is obliged to pinion his beautiful birds and to isolate them in pairs. The female Egyptian Goose watches all the fights of her mate with the interest of a boxing referee, but she never helps him, as greylag females and female cichlids do their mates; in fact, if he comes off the worse, she is always ready to go over with flying colors to the side of the winner.

Such behavior must exercise a significant effect on sexual selection, by putting a premium on the greatest possible fighting power and bellicosity of the male. Here again, a thought presents itself which engaged our attention at the end of Chapter Three. Quite probably the fighting urge of the Egyptian Goose, which often seems insane to the observer, is the result of intra-specific selection and is of absolutely no survival value. This thought is disturbing because, as we shall see later, similar considerations have to be borne in mind concerning the phylogenetic development of the human aggressive drive.

The Egyptian Goose belongs, further, among the few species in which the triumph cry can fail in its appeasing function. When two pairs of Egyptians, one pair on each side of a transparent but impenetrable fence, tease each other and work themselves up into a rage, it sometimes happens that suddenly, as though on command, the mates of each pair face each other and beat each other unmercifully. This behavior can also be induced by putting a scapegoat of the same species into the enclosure of a pair and, when the fight is in full swing, removing it as unostentatiously as possible. Then the pair indulges in an ecstatic triumph ceremony which becomes wilder and wilder, differing less and less from unritualized threatening, till suddenly the loving partners have each other by the neck and are thrashing each other hard; this always ends in the victory of the male, since he is appreciably bigger and stronger than the female. However, in the absence of a "bad neighbor," the damming of aggression in Egyptian geese never leads, as far as I know, to mate-murder in the way I have described it in cichlids.

However, in the Egyptian Goose and in most of the Sheldrake species, the main significance of the triumph ceremony lies in its functions as a lightning conductor. It is used particularly when thunderstorms threaten, that is, when both the inner mood of the animals and the external situation strongly elicit intra-specific aggression. Although the triumph ceremony, particularly in the Common Sheldrake, consists of highly differentiated, choreographically exaggerated motor patterns, it is here not dissociated from the original drives underlying the conflict, as is the case in the movement pattern of the less highly developed "greeting" of many dabbling ducks described on page 179. The triumph ceremony in sheldrakes evidently still derives its energies largely from the primal drives from whose conflict the redirected movement first developed, and it is still dependent on the simultaneous rousing of aggression and of the factors acting against it. In these species, the ceremony is subject to strong seasonal fluctuations; it is most intensive during the reproductive season, fades during the rest period, and is naturally nonexistent in young, sexually immature birds.

All this is entirely different in the Greylag and all true geese: firstly, their triumph rite is no longer the concern of the mated partners alone but has become a bond embracing the whole family and indeed whole groups of individuals. The ceremony has become entirely independent of sexual drives; it is performed throughout the year, and even tiny goslings take part in it.

The movement sequence is longer and more complicated than in all the hitherto described appeasing ceremonies. While in Cichlids, and also in many species of Sheldrake, the aggression diverted from the partner by the greeting ceremony leads to a subsequent attack on the hostile neighbor, in geese this attack precedes the affectionate greeting in a ritualized sequence of movements. In other words, it is characteristic of the triumph cry that one partner, usually the strongest member of the group, in pairs always the gander, proceeds to attack a real or apparent enemy, fights him, and then, after a more or less convincing victory, returns, greeting loudly, to his family. From this typical case, reproduced here in a diagram by Helga Fischer, the triumph ceremony derives its name.

The first part of the triumph ceremony, the attack performed with head and neck pointing obliquely forward and upward, and accompanied by a loud, raucously trumpeting fanfare, is called *rolling* (A in Fig. 4, and in Fig. 5). The second part, the return to the partner, with the neck stretched forward low along the ground, the head tilted upward, accompanied by low but rather passionate chatter, was termed *cackling* by Helga Fischer. The attitude of cackling closely resembles that of serious attack (compare 6 in Fig. 5 with 1 and 2 in the same figure and with E in Fig. 4), the only difference being that the neck and head do not point at the object of the gesture, as in threatening, but distinctly past it.

Motivation analysis, proceeding along the lines sketched in Chapter Six, shows rolling to be a behavior pattern of very complicated and conflicting motivation. The form of the movement shows a mixture of elements, including aggression, fear, and social contact. The same is true of the accompanying, unique production of sounds. In order to elicit "rolling," two entirely different sets of stimuli must be simultaneously present: those represented by the presence of a friendly cackling partner as well as those emanating from a hostile stranger. In many ways, the situation activating the rolling attack is comparable to the one releasing the critical response which we learned about on page 28. The gander, being closely tied to the spot where his mate or his young are to be found, is prevented from fleeing as effectively as the proverbially cornered

Fig. 5

rat. As with the rat, his all-out assault is the more furious and desperate, the more frightening the antagonist at whom it is launched. This is borne out by the fact that young, newly mated ganders who have not yet acquired social status are more furious and persistent in their rolling attacks than older birds possessing an assured position in the ranking order of the flock.

Unlike rolling, the second part of the triumph ceremony, cackling, is dependent on a single motivation, as the thorough analysis made by Dr. Fischer has proved beyond reasonable doubt. The expression movement with which the cackling goose turns toward the partner closely resembles the threatening gesture depicted in Fig. 4 E, and it is, indeed, only distinguishable from it by the slight deviation caused by the ritualized redirection already mentioned. Viewed in profile, this is quite imperceptible and neither man nor goose can tell whether a goose approaching another in this attitude intends to cackle with it or to launch an attack against it. In spring, when the family ties slacken and young ganders go courting, it may easily happen that one brother still continues to perform a family triumph rite with another, while he is already trying to make a strange young female an offer of pairing. This consists not in copulatory proposals but in attacks on strange geese which he performs and then hurries back, greeting, to the female of his choice. If his brother sees this from one side, he believes that the suitor wants to attack the young female, and since male members of a triumph-rite group defend each other valiantly, he rushes furiously at his brother's prospective bride and, having no tender feelings toward her for himself, thrashes her in a way which would correspond to the expression movements of his brother if he were threatening and not greeting. When the female flees in terror, her suitor finds himself in a situation of extreme embarrassment. This is not anthropomorphizing, since the objective physiological basis of every embarrassment is the conflict between opposing impulses, and the young gander is undoubtedly in just such a situation: the urge to defend the courted female is tremendously strong in a greylag gander, but equally strong is the inhibition against attacking his brother, who is still his fellow in the fraternal-triumph ceremony. We shall later see from some impressive examples how insurmountable this inhibition is.

Comparative study of other ducks and geese leaves no doubt that cackling has evolved, by way of ritualization of a redirected activity, from threatening gestures, in a manner exactly analogous to the origin of the appeasement gestures in cichlids discussed on page 169. Yet in its present form the cackling of greylag geese contains no aggressive motivation. Nor indeed does the triumph ceremony still function as an appeasement ceremony in this species. Only in a quickly traversed stage in individual development can we demonstrate, in the greeting pattern, the primal drives underlying the reorientation as well as its appeasing function. Otherwise the ontogenetic development of the triumph ceremony is obviously not a recapitulation of its evolution. Even before a gosling can stand, walk, or eat, it is perfectly able to perform the motor pattern of stretching the neck forward, simultaneously uttering a falsetto cackling. The call is at first bisyllabic, exactly like the mallard's "rab-rab" and the corresponding duckling sound. A few hours later it becomes a polysyllabic "veeveevee" whose rhythm exactly corresponds with the greeting chatter of adult geese. The stretching forward of the neck and the vee sounds are undoubtedly the preliminary stage from which, in the growing goose, the expression movements of threatening, as well as the essential second phase of the triumph ceremony, develop. We know from comparative research that, in the course of phylogenetics, greeting has evolved from the threatening by way of redirection and ritualization. But in individual development, the formally similar gesture at first has the meaning of greeting. When the gosling has accomplished the difficult and by no means undangerous task of hatching and is lying there, a little wet bundle of misery with droopingly outstretched neck, there is one reliable reaction that can promptly be elicited from it: if one bends over it and utters a few sounds in an approximately goosy tone of voice, it lifts its little head, wobbly and uncertain, stretches its neck forward, and greets. Before it is able to do anything else, the tiny wild goose greets its social surroundings!

In their meaning as an expression movement, as also in respect of the eliciting situation, the neck stretching and whispering of the little greylag resemble the greeting and not the threatening of the adult. It is, however, remarkable that, in its form, this behavior pattern is at first indistinguishable from threatening, since the characteristic sideways deviation of the outstretched neck from the direction of the partner is missing in very small goslings. It is only when they are a few weeks old, and contour feathers of the juvenile plumage are visible among the down, that the behavior alters. The young birds begin to be noticeably more aggressive toward birds of the same age belonging to other families, and, with outstretched necks and whispering, they walk up to them and attempt to bite them. But since, in such quarrels between the young birds of two families, threatening and greeting gestures are exactly the same, misunderstandings often arise and brother may bite brother. In this particular situation, one sees for the first time in ontogeny the ritualized redirection of the greeting movement: the gosling bitten by his brother or sister does not bite back but breaks into intensive whispering and neck stretching, directing this markedly past the other gosling at a more obtuse angle than is the case later on in the fully matured ceremony. The aggression-inhibiting effect of these gestures becomes particularly evident; the still aggressive brother or sister abandons the attitude at once and indulges in its turn in a greeting plainly directed past its object. The phase of development in which the triumph rite reveals such a noticeably appeasing function lasts only a few days; ritualized redirection suddenly sets in and, except in certain rare cases, henceforth prevents misunderstandings.

The function of eliminating these rather rare misunderstandings between siblings is all that remains, in ontogeny, of the original appeasing function of the triumph ceremony. In its mature state, the behavior pattern, though still retaining the external form of redirected threat, is not activated by aggression, but by the independent motivation of the greeting ceremony itself, except under abnormal circumstances of which I shall speak later. All the aggression detectable, by a thorough motivation analysis, in the triumph rite is discharged during its first, "rolling" phase, and in the direction of the hostile stranger. "Rolling" continues for a few seconds as the gander, whether victorious or not, turns away from his opponent; it ceases abruptly as he approaches his mate and, as the two meet, passionate cackling ensues with heads held close together. Fig. 5 represents the motor patterns of this procedure.

The observer familiar with the respective meanings of rolling and cackling cannot help feeling that the passion of togetherness which finds its voice in this cackling is heightened by a phenomenon of contrast, by that which physiologists call a "rebound effect." Aggression having been discharged at the hostile neighbor, tenderness toward the mate and children wells up unchecked. Conversely, the nearness of the loved ones enhances the intensity of aggression toward the intruding stranger. The family which has to be defended acts in some way like a movable territory, an interesting fact to which we shall return later on. Conversely, the presence of aggression-eliciting outsiders considerably enhances the readiness to cackle lovingly with the partner or partners in the triumph ceremony.

There is one impressive special case of the triumph ceremony in which the dual, mutually enhancing function of rolling and cackling becomes particularly apparent, although the two parts of the ceremony are not clearly separated in time but, in a manner of speaking, go off simultaneously. In autumn and winter, when many families of geese congregate in large migratory flocks, it is not the gander alone who acts as defender of the family, rushing out to attack and to return victorious, but all members of the group united by a shared triumph ceremony set forth together to drive away another family group. Every goose is obviously torn between the conflicting urges to roll in the direction of the enemy and to cackle with the next member of the family; one can actually see the necks swinging to and fro between the two alternative directions. Finally, all the members of the family stand roughly parallel to each other, pointing threatening necks at the hostile group, while simultaneously trying to keep their heads as close together, cheek by cheek, as postulated by the rite of cackling. The result is the formation of a closed wedge-shaped phalanx of converging necks. Viewed from the front, this presents a spectacle which, together with the mixed rolling and cackling accompanying it, intimidates the enemy the more effectively, the greater the number of group members participating in this "ceremony of converging necks," as I termed it many years ago. Helga Fischer calls it, briefly and graphically, the "roll-cackle."

Discriminative aggression toward strangers and the bond between the members of a group enhance each other. The opposition of "we" and "they" can unite some widely contrasting units. Confronted with present-day China, the United States and the Soviet Union occasionally seem to feel as "we." The same phenomenon, which incidentally has some of the earmarks of war, can be studied in the roll-cackle ceremony of greylag geese. In autumn and winter it occasionally happens that flocks of geese, consisting of several families, come back from the breeding colonies which we settled some miles away on neighboring lakes, when the number of birds on our Ess-See had become excessive. Faced with these utter strangers, the otherwise mutually hostile families of geese on our lake unite in one collective phalanx of converging necks, and attempt to drive away the intruders, who, in turn, form another phalanx and usually stand their ground, provided they are numerous enough.

In all these cases the triumph rite performs a function subtly different from that of the primal appeasing ceremony from which it evolved. Though the external form of redirected threat is still all there in the stretching of the neck past the cackling partner, the latter has long ceased to arouse any aggression which needs to be redirected or which can be exploited to increase the intensity of attack against the neighbor, as in the case of the nest-relief ceremony of cichlids (page 172). Hence the temporal sequence of the redirected movement and the attack against the hostile stranger is inverted: in cichlids, the attack follows upon the redirected movement, in geese it precedes it. Yet the whole ceremony has a similar effect on the behavior of the individuals participating in it. It still holds them together and enables them to stand by each other against a hostile world. The principle of the bond formed by having something in common which has to be defended against outsiders remains the same, from cichlids defending a common territory or brood, right up to scientists defending a common opinion and—most dangerous of all—fanatics defending a common ideology. In all these cases, aggression is necessary to enhance the bond. What is so new, and indeed hope-inspiring, about the triumph ceremony is that the bond it forms is so largely independent of aggression. Geese held together by a shared triumph ceremony stay together, irrespective of whether or not they have young or territory to defend, are surrounded by hostile fellow members of the species or are all on their own. They perform their beautiful rite just as intensely on meeting again after a long separation as they would after the most glorious victory in battle.

However, the great marvel of the triumph ceremony and one that inspires even the most objective observer with human sympathy is the enduring and personal nature of the bond by which it unites the individuals participating in it.

The group embraced by the triumph ceremony is remarkably exclusive. The newly hatched bird enjoys the birthright of group membership and is accepted "unquestioningly" even when it is not a goose but an experimentally introduced changeling, such as a Muscovy Duck. After only a few days, the parents have learned to know their children and the children their parents, and from now on they are not prepared to perform the triumph rite with any other geese.

If one makes the rather cruel experiment of transplanting a gosling into a strange family, it will be found that the later it is removed from its family the more difficulty the poor baby will have in finding acceptance by the strangers' triumph community. The baby is afraid of the strangers, and the more fear it shows the more they are likely to attack it.

It is an unforgettable experience to hand-rear a gosling from the moment of its hatching. One cannot help feeling moved by its childlike trust when it stretches forward its tiny neck and whispers its little greeting to the first living being that approaches it, "taking it for granted" that this must be one of its parents.

Only in one other situation does a greylag ever offer its triumph rite, and with it permanent love and friendship, to a complete stranger: that is when a young male suddenly falls in love (without quotation marks!) with a strange young female. These first proposals take place at a time when last year's young birds are obliged to leave their parents, who are now getting ready for the new brood. At this moment family bonds necessarily loosen but they are never completely broken.

With geese, much more so than in the duck species already discussed, the triumph rite is bound up with personal recognition of the partner. Ducks, too, palaver only with acquaintances, but the bond knitted by this ceremony between the participating individuals is not so firm, nor is the group membership so difficult to acquire as in geese. If an individual goose has recently flown into a colony, or has been introduced into a group by the owner, it may literally take years before it is accepted into one of the goose groups bound by the triumph rite.

The stranger may more easily acquire a partner by suddenly falling in love, and deviously achieve membership of a larger triumph-rite group by founding a family. Apart from the special cases in which a goose has fallen in love and found its love reciprocated, or has been born into the family group, the triumph ceremony is the more intensive and the bond uniting the partners the firmer, the longer the animals have known each other. All other things being equal, one can say that the strength of the triumph-rite bond is proportional to the duration of the friendship of the partners, or one might even say that a triumph rite always develops when companionship between two or more geese has lasted over a prolonged period of time.

In the early spring, when the older goose pairs are concerned with breeding, and the many young one- and two-year-old geese with love, numerous unpaired geese of different ages are always left over as erotically unemployed wallflowers, and these then join together in groups of varying sizes. We call these groups the nonbreeders, though the expression is not an accurate one since the many young but firmly paired couples have not yet begun to breed either. In such nonbreeding groups, firm triumph-rite bonds can develop which have nothing whatever to do with sexuality. Also if two lonely geese are dependent on each other's company a nonbreeding association between a male and a female may occur. This actually happened at our station, when an old widowed goose returned from our branch settlement on the Ammersee and joined a recently widowed gander in Seewiesen. I believed that pair formation was imminent, but Helga Fischer thought from the beginning that it was only a typical nonbreeding triumph rite such as sometimes unites an adult male with an adult female.

Contrary to popular opinion, there are true friendships between male and female which have nothing to do with love, though naturally love may spring from them. If wild-goose breeders want to pair two birds which fail to respond to each other, they often transfer them together to another zoo or water-fowl collection where, as newcomers, they are so unpopular that they are dependent on each other for company. In this way, one can force the formation of a nonbreeding triumph ceremony, in the hope that pair formation will ensue. However, all too often I have found that such forced ties soon dissolve when the pair returns to its old environment.

The relation between the triumph rite and sexuality, the true copulatory drive, is not easy to understand and in any case it is only a loose one, for everything purely sexual plays a very subordinate role in the life of the wild goose. The bond that holds a goose pair together for life is the triumph ceremony and not the sexual relation between mates. The presence of a strong triumph-rite bond between two individuals "paves the way," that is furthers to a certain extent the development of sexual relations. When two geese—or two ganders—have been united for a long time by this ceremony, they usually try to copulate in the end. But conversely, the copulatory relations that often occur in year-old, sexually immature young birds do not appear to be favorable for the development of a triumph-rite bond. Young geese can often be seen making copulatory movements, but these are no forecast of a later pairing.

On the other hand, the least sign that a triumph-cry proposal from a young gander is finding response in the female means that in all probability the two will become a pair. Such tender relations in which copulatory reactions play no part appear to dissolve completely in summer and autumn, but in the following spring, when the young geese start courting seriously, they often return to their young love of the previous year. The loose and rather one-sided relations between the triumph rite and copulation in geese have far-reaching analogies with those between falling in love and physical sexual reaction patterns in man. The "purest" love leads by way of the greatest tenderness to a physical approach which, however, is in no way the essential part of this bond; and conversely, those stimulus situations and partners which release the strongest copulatory drives are certainly not unconditionally the ones that produce the most intense falling in love. In the Greylag Goose, the two function cycles can dissociate themselves as completely from each other and make themselves as independent as they can in man, though they undoubtedly belong together "normally" and must relate to one and the same partner if they are to fulfill their survival value.

The concept "normal" is one of the most difficult things to define in the whole of biology, but at the same time it is unfortunately as indispensable as its counterpart, the concept "pathological." My friend Bernhard Hellmann, when confronted with something particularly bizarre and unaccountable in the structure or behavior of an animal, used to ask the apparently naïve question, "Is this how the constructor meant it to be?" In fact, the only way to assess "normal" structure and function is by demonstrating that these are the ones which, under selection pressure of their survival value, have evolved in this and in no other form. Unfortunately, this definition leaves out of consideration everything which has become so and not otherwise by pure chance, though such chance products of evolution need not necessarily fall into the category of the pathological. However, by normal we understand not the average taken from all the single cases observed, but rather the *type* constructed by evolution, which for obvious reasons is seldom to be found in a pure form; nevertheless we need this purely ideal conception of a type in order to be able to conceive the deviations from it. The zoology textbook cannot do more than describe a perfectly intact, ideal butterfly as the representative of its species, a butterfly that never exists exactly in this form because, of all the specimens found in collections, every one is in some way malformed or damaged.

We are equally unable to assess the ideal construction of "normal" behavior in the Greylag Goose or in any other species, a behavior which would occur only if absolutely no interference had worked on the animal and which exists no more than does the ideal type of butterfly. People of insight *see* the ideal type of a structure or behavior, that is, they are able to separate the essentials of type from the background of little accidental imperfections. When my teacher, Oskar Heinroth, described in his now classic work on the Anatidae (1905) the lifelong, unconditional marital fidelity of the Greylag Goose as its "normal" behavior, he had correctly abstracted the ideal, interference-free type, though he could never have observed it in reality, since the life of a goose may last half a century and its marriage only two years less. Nevertheless, his assertion is justified and the type described by him is as useful as a norm deduced from the average of many single cases would be useless. Shortly before writing this chapter, I was working with Helga Fischer over her goose records, and in spite of what I have just said I evidently showed disappointment that Heinroth's type of perfect goose marriage, faithful unto death, was so rarely to be found among our many geese. Whereupon Helga, exasperated by my disappointment, made the immortal remark, "What do you expect? After all, geese are only human!"

In wild geese—and this has also been demonstrated in free-living ones—there are very wide divergences from the norm of marital behavior; and of these, one which frequently occurs is of special interest because, though strictly condemned in some human cultures, in geese, surprisingly, it is not harmful to survival; I mean the bond between two males. There are no obvious qualitative differences either in appearance or behavior between the two sexes; only one ceremony in pair formation, the so-called angle-neck, is essentially different in the two sexes and presupposes that the prospective partners are unacquainted before pairing and therefore a little afraid of each other. If this rite is skipped, a gander may possibly make his triumph-rite proposal to another male instead of to a female. This happens frequently, and not only when, owing to close confinement, all the geese concerned know each other too well. At my former research station, in Buldern, Westphalia, where we were obliged to keep our geese on a relatively small pond, this happened so often that for a long time we thought that in the pair formation of greylags, heterosexual partners only came together by trial and error. It was only later that we discovered the function of the angle-neck ceremony, but this need not be discussed here.

If such a young gander makes his triumph-rite proposal to another male and is accepted, they each find, in respect of this all-important ceremony, a far better partner and companion than they would have done in a female. Since intra-specific aggression is far stronger in ganders than in geese, the inclination to perform the triumph ceremony is also stronger and the two friends stimulate each other to acts of courage. No pair of opposite sex can compete with them; thus such gander pairs always attain very high, if not the highest, places in the ranking order of the colony. They keep together for life just as faithfully as a pair of heterosexual individuals. When we separated our oldest gander pair, Max and Kopfschlitz, and sent Max to our branch colony on the Ampersee near Fürstenfeldbruck, after a year of mourning both ganders paired with female geese and bred successfully, but when we fetched Max back—without his wife and children, whom we could not catch—Kopfschlitz immediately deserted his family and returned to him. Kopfschlitz' wife and sons seemed to understand the situation, and they made furious but unsuccessful attempts to drive Max away. At the time of writing the two ganders are still united and Kopfschlitz' abandoned wife waddles slowly after them at a measured distance.

Our concepts of "homosexuality" are very wide and very ill defined. The effeminate, painted youth in the bar and the hero of a Greek epic are both considered homosexual, although the behavior of the former comes close to that of the opposite sex, whereas that of the latter is the behavior of the true superman, and it is only in his choice of a sexual partner that he deviates from the normal. Our "homosexual" ganders come under the second category, and their erring is far more "forgivable" than that of Achilles and Patroclus because male and female differ much less in geese than in human beings. Moreover, their behavior is far less "animal" than that of most human homosexuals, for they seldom if ever copulate or perform substitute actions. In spring, they can be seen celebrating with graceful neck dipping the ceremony of precopulatory display. The poet Hölderlin put this behavior, in swans, into verse. If, after the ritual, the ganders want to copulate each tries to mount the other and neither thinks of laying himself flat on the water, after the manner of the female. Developments being held up that way, they may become frustrated and angry with each other, but just as frequently they give up their attempt with no signs of emotion or disappointment. In a manner of speaking, each takes the other for a female, but the fact that "she" is a little frigid and simply will not be mated scarcely interferes with the great love. With the advance of spring, the ganders gradually learn that they cannot copulate and they no longer attempt to mount each other, but during the winter they forget their inability and next spring, with fresh hope, they attempt to mate again.

Often, but not always, the mating drive of two ganders

linked by the triumph rite finds outlet in another direction: probably their social superiority, attained by their united strength, has a particularly strong attraction for unpaired females; at all events, a goose will soon be found to follow the two heroes at a respectful distance, since, as subsequent events show, she has fallen in love with one of them. This young female at first stands or swims beside them, but when the ganders make their unsuccessful attempts at copulation, she soon cunningly learns to push herself between them in an attitude of readiness at the critical moment when the male of her choice is trying to mount the other. She always offers herself to the same gander, which then regularly mates her but immediately afterward turns to his friend and addresses the ceremony of postcopulatory display to him. "It was really you I was thinking of, all the time!" The second gander often takes part in the epilogue. In a particular, recorded case the goose did not follow the two ganders everywhere, but waited in a certain corner of the pond for her lover till midday, when geese are particularly ready to copulate; he swam quickly toward her, but immediately after mating her he flew straight back across the pond to his friend—a most unchivalrous act toward the female. But she never seemed to take offense.

In the gander, such a loveless copulatory relationship with a female may gradually become a habit, while in the goose concerned there is always a latent readiness to join in his triumph ceremony. The longer they know each other, the smaller becomes the distance between the goose and the two ganders, and gradually the second male, too, becomes used to her presence. Very gradually she begins to join in the triumph ceremony of the two friends, first shyly and later with increasing confidence, until the ganders are quite used to her participation. So the goose, by the devious way of long, long acquaintance, changes from the status of a more or less unwanted appendage of one of the ganders to an equal member of the triumph-rite community.

This long procedure may be shortened if the goose, who at first receives no help from anybody in the defense of a nest territory, manages nevertheless to acquire a nesting place for herself and to breed. Then the two ganders may discover her and adopt her, either during incubation or after the hatching of the young. In reality they are adopting the brood, and they merely tolerate the fact that the mother joins in when they perform their triumph rite with the adopted children, who are of course the offspring of one of the two ganders. Standing watch over the nest and guiding the children are, as Heinroth has written, the climax in the life of a gander and are evidently charged with more feeling and emotion than precopulatory display and copulation; as a result, they build a better bridge than copulation for a closer acquaintance of the individuals and the formation of a shared triumph ceremony.

Finally, by whichever of these different means it may be, after some years a real triangular marriage always results; the second gander also begins to mate the goose and all three birds enact the precopulatory and postcopulatory displays together. The most remarkable thing about such triangular marriages, of which we were able to observe a considerable number, is their biological success: their members are always at the top of the ranking order of their colony, they are never driven out of their nesting territory, and they raise, year by year, a considerable number of goslings. Thus one cannot regard the "homosexual" triumph bond of two ganders as something pathological, the less so since it also occurs in wild geese in the natural state. Peter Scott observed in wild pink-footed geese in Iceland a considerable percentage of families consisting of two males and a female. The biological advantage of double defense by the fathers was even more evident there than in our local geese, which are comparatively protected against predators.

I have described how a new member can be received into the exclusive circle of a triumph community by way of long habituation, but it still remains for me to set forth how such a bond may develop explosively, almost before one has realized it, joining two individuals together for life. We say then, without quotation marks, that the two have fallen in love. The expression aptly describes the suddenness of the process.

The overt behavior of females, and also of very young males, is not as conspicuously changed by their falling in love as is that of adult ganders. Not that the consequences are less momentous for them—quite the contrary; but all their actions are restrained by a natural shyness that is very appealing. Fully matured ganders advertise their new love by all the means at their disposal, and it is really quite amazing what changes can be wrought in the outer appearance of a bird lacking special organs of display. A fish can make its colors glow in sudden iridescence, it can unfold imposing fins, a peacock can raise and rustle his wonderful tail coverts, a man can dress up in his best clothes in order to appear as different as possible from his common everyday self. A gander can do none of these things, and yet it has happened to me that I simply failed to recognize a well-known individual after he had suddenly fallen in love. The tension of all muscles is increased, which gives the bird a curious stance of pride and alertness and changes the contours of his body considerably. Every single movement, walking, swimming, and flying, even preening or turning the head, is performed with an excessive expenditure of energy. Taking wing, a procedure which otherwise requires resolution and effort, comes as easily to a gander in love as to a hummingbird; he will fly even over very short distances which are really not worth the effort of taking off and landing again, and which every goose in its right mind would cover walking or swimming; he enjoys sudden acceleration and deceleration rather like a boy on a motorcycle. He particularly enjoys flying swiftly toward his bride, putting on the brakes at the latest possible moment and landing in a flurry of beating wings right in front of her, to deliver a triumph ceremony. Most of all he enjoys bullying and beating up other ganders while his lady is looking.

A young female that has fallen in love never tries to force her company on the object of her passion. She never follows him directly when he walks away; she merely turns up, as if by chance, in places where she knows he can often be found. If the gander does indeed court her, she does not react, for a considerable time, by attitude or gesture. It is only the play of her eyes which tells the male how his courtship is received. Though she never looks directly at him and pretends, instead, to be looking at something else, she still watches his antics with the greatest interest. As she tries to do this without noticeably turning her head, she has to squint at him out of the corner of her eyes just like a girl flirting.

As with human beings, Cupid's dart occasionally chances to hit one individual only. According to our records this happens to males more frequently than to females. In this, however, they could easily be wrong owing to the fact mentioned above that young females are much more restrained in the expression of their love, so that their secret passion is easily overlooked. In males, a one-sided love quite frequently achieves a happy ending, even if the gander does not for a long time find the desired response. The male is able to force his attention on the female, to follow her around pertinaciously, and, above all, to spoil the chances of all rivals he is capable of vanquishing. A goose finds it difficult to love a male who is too patently afraid of another, so a gander, provided he persists long enough in his advances, usually succeeds in getting the female so accustomed to his constant offers of a triumph ceremony that she finally joins in.

Permanently unrequited unhappy love is the fate of those who become attached to individuals whose affections are already absorbed in a happy mating. Ganders, in this case, very soon desist from their unsuccessful courtship; at least I have never known one to pursue a well-mated female for years. The record of a very tame hand-reared female, on the other hand, shows that her faithful love for a happily mated gander persisted for more than four years. She never openly followed him and his family, she always "just happened" to be where he was. Every year she proved her own fidelity and, incidentally, that of the gander to his female, by laying an infertile clutch of eggs.

Fidelity over the joint performance of the triumph ceremony and fidelity over copulation are correlated to each other in a peculiar way which is slightly different in males and females. In the ideal "normal" case, in which everything proceeds without disturbance, that is to say when two handsome and healthy geese are betrothed in the first, or at the latest in the second, year of their life, when neither of the two kills itself by flying against a high-tension cable, or gets lost in a fog, or is eaten by a fox, or is infected with nematodes, etc., etc., both male and female remain absolutely faithful to each other forever after, both in respect of the triumph ceremony and of copulation. If a cruel fate sunders the bond of love at an early stage, both gander and goose are able to develop a new triumph ceremony with another partner; the sooner the separation from the first mate took place the easier it is. Except for cases in which this happened at the very beginning of an attachment, the loss of his first love has a queer permanent influence on the behavior of a gander: he is not as faithful to his new wife in regard to copulation as he would be if he had married the "first woman in his life." In respect of the triumph ceremony his behavior is not at all disturbed; his responses concerning the defense of the nest and the family are in no way impaired; he is a model husband and father—except that he will copulate, whenever opportunity offers, with strange females. Funnily enough, he is the more prone to such lapses the farther he happens to be away from his wife, or from the nest while she is incubating. If a strange female comes near his nesting site, he will drive her away quite furiously, but he is ready to copulate with her if he meets her in some other place. The anthropomorphic observer tends to interpret this type of behavior as an attempt, on the part of the gander, to keep his affairs secret from his wife. This, of course, greatly overrates the bird's mental powers. The real explanation is that, near his nest, he reacts to the strange female with territorial defense just as he would to any other trespassing goose which was not a member of his family. Away from his territory, on the other hand, his defense responses do not suppress the sexual one which the stranger elicits on neutral ground. A gander may have affairs of long duration, regularly meeting a female other than his wife in a "secret" place and copulating with her. However, she is his partner only in copulation; he never accompanies her when she is walking and he never gives even the slightest hint of a triumph ceremony in her presence. In this respect he remains absolutely faithful to his wife. Nor does he guard the strange female's nest; should she happen to be successful in acquiring a nest site and in rearing a family, she must do so unaided by the gander. He does not love her in the least.

The strange female, for her part, is indeed in love with the gander; she attempts to stay near him in the restrained, modest way already described, and she would at once accept his offer of a triumph ceremony should he make one. In the female greylag geese, the urge to copulate and the urge to give the triumph ceremony cannot be dissociated and discharged with two separate objects as so often happens in ganders. Also, the female goose finds it much more difficult than the male to develop a triumph ceremony with a new partner after having lost the old one. This is particularly the case with geese who have been widowed for the first time. The more often they have lost their mate or have been left by him—which also happens on occasion—the easier they find it to form a new triumph ceremony. However, the more often such a new bond is formed, the weaker and less durable it becomes. A goose who, through no fault of her own, has become widowed or divorced a number of times, shows a type of sexual behavior vastly different from that of a goose mating for the first time. Being sexually much more active and not inhibited by the "maidenly restraint" so characteristic of virgin geese, she is equally ready to enter a new triumph-ceremony relationship and to copulate. Such a goose can become a real *femme fatale*. She actively provokes the courtship of a younger gander offering her a permanent triumph ceremony, only to make him deeply unhappy a few weeks later by leaving him for a new lover. The record of the marriages contracted by our oldest greylag, Ada, exceeds that of a film actress and differs from it by the quite untypical end: in the ninth year of her life she found a *grande passion* in the person of a hitherto unmated gander of nearly her own age and remained faithful to him for the rest of her life. She raised a brood every year and died in 1963 from being egg-bound. Her life history will be told in another book.

The longer a pair has been happily mated and the nearer the history of their marriage comes to the ideal case described above, the more difficult it becomes for a widowed bird to form a new triumph ceremony. As I have shown, it is harder for the female than for the male. Heinroth recorded some cases in which widowed female geese never mated again and stayed sexually inactive for the rest of their lives. We never observed a similar fidelity in ganders. One bird, widowed late in life, "mourned" for exactly one year, then started having copulatory affairs with a considerable number of unmated females and finally achieved a new triumph-ceremony bond by the indirect procedure described on page 198. There is one exceptional case in which a female goose who had been well mated and had led an exemplary married life for some years, on the loss of her husband immediately entered a new triumph ceremony with another gander and lived happily with him forever after. Our explanation that something must have been wrong with the first marriage sounds too much like begging the question.

Such exceptions, however, are so extremely rare that in order to convey the correct idea of the general durability and strength of the bond, by which the triumph ceremony embraces the mates of a pair and, indeed, all the members of a family, I should perhaps have done better not to mention them. Not only in the ideal case of undisturbed pair formation, but on the average of all cases recorded, the triumph ceremony constitutes the most important structural element on which the social life of geese is built up. It is, so to speak, the leitmotiv of goose life. As a slight undertone, it is present in all their daily activities. Its lowest intensity, the ordinary low cackling, so aptly translated by Selma Lagerlöf by the words "here I am, where are you?," can be heard almost permanently at short intervals while they are grazing or walking slowly; it swells up when two slightly hostile families come near each other; it is silent only when they are seriously alarmed, when they are fleeing, or when they are covering great distances on the wing. The moment, however, that whatever occupation interrupts or prevents the continuous communing of the partners is happily past, the greeting ceremony breaks out again, the louder and more intensely the longer it has been suppressed. On the reunion of partners that have been separated for an appreciable time, say for a few hours, the full-blown ceremony, in its highest intensity, is triumphantly performed.

It would seem that the partners to a triumph-ceremony group have to reassure each other all day long and at every opportunity that they do indeed belong together, forming an independent social entity. In reality the relationship of cause and effect is the other way around. The triumph ceremony is not caused by love and friendship between certain individuals, it is not "the expression of" these feelings, quite the contrary, the ceremony itself is instrumental in keeping the group members together.

I have already explained that all instinctive motor patterns possess their own spontaneity, their own action-specific production of endogenous stimulation whose quantity is quite exactly calculated to meet the demand which is to be expected in the organism's normal environment. The more frequently a certain motor pattern is normally used, the greater its supply. Mice must gnaw a lot, hens must peck a lot, and squirrels jump a lot under natural conditions, and it is not immediately apparent that this "must" comes more from an internal urge than from external releasing stimulation. If such stimulation is withheld experimentally, even for a short time, it becomes quite clear that environmental stimuli determine only when and where this instinctive movement is performed, and that their absence does not prevent its performance. The mouse gnaws interminably at the most ungnawable substitute objects; the hen pecks and pecks, for lack of better objects, at the feathers of her fellow prisoners; the squirrel jumps incessantly, turning somersaults in its narrow cage. An analogous physiological process forces the Greylag Goose to perform the triumph ceremony, and if circumstances prevent it from doing so, the wretched bird becomes a pathological parody of its normal self. The goose cannot even discharge the dammed-up urge at a substitute object, as the mouse, the hen, and the squirrel can.

A greylag which does not have any partner with whom to perform the triumph ceremony is permanently depressed and sits or stands about moping. Yerkes once said that *one* chimpanzee is not a chimpanzee at all, and a similar statement is even more emphatically true of a greylag. Even within a populous colony of fellow members of its own species, a single individual not belonging to any triumph-ceremony group suffers severely from its loneliness. If one produces such a sad state of affairs experimentally, by rearing a goose in complete isolation from its own species, one can regularly observe a number of characteristic disturbances in the unhappy creature's response to its environment in general and to its social environment in particular. It is a matter of extreme interest that these disturbances are in many ways analogous to those observed by René Spitz in hospitalized children who were deprived of sufficient contact with adult human beings. What gets more damaged than all other functions is the faculty of dealing actively with new environmental situations. So far from actively striving for new and rewarding stimuli and exploring their environment, these poor creatures try to avoid all forms of stimulation and act exactly as if these were painful to them. The attitude of lying prone in their cots with their faces turned toward the wall is symptomatic of damage done by hospitalization. Social contact is especially shunned, and even children who are only slightly damaged in this way never look each other or, for that matter, anybody else straight in the face. When put together in the same pen, two young geese who had been experimentally crippled in the manner just described, avoided each other meticulously and regularly sat in opposite corners with their backs turned. René Spitz, to whom we were able to demonstrate this experiment, was struck by the amazing analogy between the behavior of these geese and that of the children he had studied in orphan asylums. Unlike human children, young greylags slowly recover from the damage wrought by deprivation of social contact during infancy. We do not know yet how complete this recovery will be, as restoring to normal takes a very long time. As yet none of our experimental geese has mated.

Even more dramatic than the disturbances caused by preventing a goose from forming normal social contacts are the consequences of sundering the bond of a triumph ceremony after it has been firmly established. While early deprivation of social contact creates an entirely artificial situation, the sudden and shattering loss of the triumph-ceremony partner must be a frequent happening in the dangerous life of wild geese. The first response to the disappearance of the partner consists in the anxious attempt to find him again. The goose moves about restlessly by day and night, flying great distances and visiting all places where the partner might be found, uttering all the time the penetrating trisyllabic long-distance call. Often our attention is drawn, by this behavior, to the fact that one of our geese is missing. The searching expeditions are extended farther and farther, and quite often the searcher himself gets lost, or succumbs to an accident. From the moment a goose realizes that the partner is missing, it loses all courage and flees even from the youngest and weakest geese. As its condition quickly becomes known to all the members of the colony, the lonely goose rapidly sinks to the lowest step in the ranking order. The threshold of flight-eliciting stimulation becomes lowered not only in respect to the attack of conspecifics, but to all other fear-inspiring circumstances as well; the goose can become extremely shy, reluctant to approach human beings and to come to the feeding place; the bird also develops a tendency to panic which further increases its "accident-proneness."

All this applies to half-tame geese reared by their own parents. In hand-reared birds, the loss of the triumph-ceremony partner can have apparently opposite effects. Geese which, through years of happy marriage, have not shown the least attachment to their former foster parents, may quite suddenly become strongly reattached to them after being widowed. This is what the gander Kopfschlitz did, after we had exiled friend Max, as told on page 196. After our attention had been called to this phenomenon, we repeatedly observed that geese, after having lost their partners, took up again their long-neglected connection with parents and siblings. Probably it is due to the same recrudescence of family ties that geese which we had settled, as adult birds, on neighboring lakes to form new colonies, regularly returned to Seewiesen if they lost their mates.

All the objectively observable characteristics of the goose's behavior on losing its mate are roughly identical with those accompanying human grief. This applies particularly to the phenomena observable in the sympathetic nervous system. John Bowlby, in his study of infant grief, has given an equally convincing and moving description of this primal grieving, and it is almost incredible how detailed are the analogies we find here in human beings and in birds. Just as in the human face, it is the neighborhood of the eyes that in geese bears the permanent marks of deep grief. The lowering of the tonus in the sympathicus causes the eye to sink back deeply in its socket and, at the same time, decreases the tension of the outer facial muscles supporting the eye region from below. Both factors contribute to the formation of a fold of loose skin below the eye which as early as in the ancient Greek mask of tragedy had become the conventionalized expression of grief. My dear old greylag Ada, several times a widow, was particularly easy to recognize because of the grief-marked expression of her eyes. A knowledgeable visitor who knew nothing about Ada's history standing beside me at the lake suddenly pointed her out among many geese, saying, "That goose must have been through a lot!"

It is on principle impossible to make any scientifically legitimate assertion about the subjective experiences of animals. The central nervous system of animals is constructed differently from ours, and the physiological processes in it are also different from what happens in our brain. These qualitative differences are sufficient to make us conclude that whatever subjective phenomena may correspond to neural processes in animals must be considerably different from what we, ourselves, experience. However, similarities and analogies in the nervous processes of animals and men are sufficiently great to justify the conclusion that higher animals do indeed have subjective experiences which are qualitatively different from but in essence akin to our own. We are convinced that animals do have emotions, though we shall never be able to say exactly what these emotions are. My teacher Heinroth, who was most careful to describe animal behavior as objectively as possible, was often accused by animal lovers of misrepresenting living creatures as being machines, because of his mechanistic interpretations of behavior. To such aspersions he used to answer: "Quite the contrary, I regard animals as very emotional people with very little intelligence!" We cannot know what a gander is feeling when he stands about displaying all the symptoms of human grief on the loss of his mate, or when he rushes at her in an ecstasy of triumph calling on finding her again. But we cannot help feeling that whatever he may experience is closely akin to our own emotions in an analogous situation.

Considered objectively, the whole behavior of a wild goose deprived of its triumph bond is highly similar to that of a highly territorial animal if it is uprooted from its home environment and put in a strange one. We see the same desperate searching and the same ebbing of courage all the time the animal is looking for its old surroundings. The greylag goose behaves toward her triumph partner just as a resident animal does toward the center of its territory, being the more tied to it the longer she has known it. Near the center of the territory, not only intra-specific aggression but many other autonomous activities of the species reach their highest intensity. Monika Meyr-Holzapfel called the partner that is a personal friend "the animal with the home valency," and with this term, avoiding all anthropomorphic subjectivity, she has apprehended the fullness of the emotional values pertaining to the true friend.

Poet and psychoanalyst alike have long known how close love and hate are, and we know that in human beings also the object of love is nearly always, in an ambivalent way, an object of aggression too. The triumph ceremony of geese—and this cannot be stressed too often—is at most an extremely simplified model of human friendship, but it shows significantly how such an ambivalence can arise. Though in the greylag the second act of the ceremony, the friendly turning toward the partner, normally contains almost no more aggression, the ceremony as a whole, particularly the first part with its accompaniment of "rolling," contains a certain measure of autochthonous aggression directed, if only latently, toward the well-loved friend and partner. We know that this is so, not only from the phylogenetic considerations discussed in this chapter, but also from observations of exceptional cases throwing light on the interaction of the original aggression and the now autonomous triumph-ceremony activation.

Our oldest snow gander, Paulchen, paired in his second year with a snow goose of the same age but kept at the same time a triumph bond with a second snow gander, Schneerot, with whom he lived in brotherly affection. Now snow geese have a habit, common in perching and diving ducks but uncommon in geese, of raping strange females, particularly when they are incubating. The following year, when Paulchen's wife built her nest, laid her eggs, and was sitting on them, an interesting but nasty situation arose: Schneerot raped her persistently and brutally and Paulchen made not the slightest attempt to do anything about it. When Schneerot came to the nest and set upon the female, Paulchen rushed at him furiously but at the last moment swerved past him and attacked any harmless nearby object, for instance our photographer who was filming the scene. Never before was the power of redirection, fixed by ritualization, brought home to me so forcibly: Paulchen wanted to attack Schneerot but could not, because the fixed path of the ritualized movement pattern directed him as firmly past the object of his anger as the points of a railway line direct a locomotive into a siding.

As the behavior of this gander conclusively shows, aggression-eliciting stimuli, if proceeding from the partner, release only the triumph cry and not attack. In the Snow Goose the two acts of this ceremony—the first more aggressive and directed outward, the second an almost entirely socially motivated turning toward the partner—are not so markedly divided as they are in the Greylag. A snow goose, particularly in its triumph ceremony, seems to be more imbued with aggression than the friendly greylag, and its triumph cry is more primitive than that of its gray relative. So in the abnormal case just described we may find a behavior corresponding in the mechanisms of its impulses with the primal redirected attack glancing past the partner, such as we have already learned about in cichlid fishes. The Freudian concept of regression is applicable here.

A rather different kind of regression may cause certain changes in the second, less aggressive phase of the triumph cry of the Greylag which show the original participation of the aggression drive. The highly dramatic scene is enacted only when two strong ganders have formed a triumph bond. Since even the most belligerent goose is inferior in strength to the smallest gander, no normal goose pair can ever win a fight against two such friends. Therefore such ganders regularly stand high in the ranking order of the colony. Now with age and with long tenure of high rank "self-assurance," that is assurance of victory, increases, and with it intensity of aggression.

Since at the same time the intensity of the triumph ceremony increases, as we have seen on page 192, with the degree of acquaintance of the partners, that is with the duration of their association, it is understandable that the ceremony of alliance in such a gander couple reaches grades of intensity never attained by pairs of unlike sex. The ganders Max and Kopfschlitz, "married" for the last nine years, are recognizable from afar by the wild enthusiasm of their triumph ceremony.

Now the triumph ceremony of such ganders sometimes increases beyond all measure, to the pitch of ecstasy, and then something very remarkable and sinister happens: the sounds become increasingly stronger, quicker, and more concentrated, the necks more and more horizontal till the upright attitude typical of the ceremony is lost and the angle between the line of the redirected movement and the line pointing directly toward the partner becomes smaller and smaller. In other words, with extreme increase of intensity the ritualized ceremony loses more and more those characteristics which differentiate it from its unritualized prototype. Thus it regresses, in the Freudian sense, to a phylogenetically earlier primitive condition. J. Nicolai was the first to discover this "deritualization" in bullfinches. In these birds the greeting ceremony of the female has arisen, like the triumph ceremony of geese, from ritualization of primally threatening gestures. If we increase the sexual drives of a female bullfinch by submitting her to a long period of solitary confinement, and afterwards put her with a male, she will pursue him with greeting gestures whose character is the more aggressive the more the sexual drives have been dammed. The tumult of such ecstatic love-hate can, in the gander couple, halt at any stage and subside; then follows a triumph ceremony still very excited but normally ending in quiet, tender cackling even if the gestures still have the expression of furious aggression. When one sees such an exhibition of fervent love for the first time, without yet knowing anything about the phenomena above described, one experiences a certain feeling of uneasiness: involuntarily one remembers such expressions as "I love you so much I could eat you" and one recalls what Freud so often stressed, that colloquialisms often reflect deepest psychological associations.

However, in our goose records of the last ten years we have only three cases in which the deritualization of the triumph ceremony, rising to highest ecstasy, did not subside. In these cases there occurs something irrevocable and of great consequence for the future life of the individuals: the threatening and fighting attitudes of the two ganders assume purer and purer forms, excitement rises to boiling point, and suddenly the two friends seize each other by the neck and beat each other with resounding blows of their wings, which are armed with hard little horns at the shoulder. Their grim fight can be heard from afar. While an ordinary fight between two ganders, induced by rivalry for status, for a female, or for a nesting place, seldom lasts more than a few seconds and never more than a minute, we registered in one of three fights between triumph-cry partners a full quarter of an hour after we had rushed to the scene, alarmed from afar by the sound of battle.

The embittered fury of these fights is only partly explained by the fact that the opponents know each other so well that they are less afraid of each other than of strangers. In human beings, too, the particular horror of marital quarrels springs only partly from this source. I am much more inclined to believe that in every case of genuine love there is such a high measure of latent aggression, normally obscured by the bond, that on the rupturing of this bond the horrible phenomenon known as hate makes its appearance. There is no love without aggression, but there is no hate without love!

The victor never pursues the vanquished, and we have never known a second fight to take place between the two ganders. On the contrary, they avoid each other meticulously from thenceforth, and when our big flock of geese is grazing on the marsh, the former friends that have fallen out with each other are always to be found at opposite sides of the periphery. If by chance, not having noticed each other in time, or owing to our experiments, they do come near each other, they show the most remarkable behavior that I have ever seen in animals, and I hesitate to describe it for fear of being accused of anthropomorphizing. The ganders are embarrassed! They cannot look at each other. Their glances dart hither and thither, magnetically attracted to the object of their love and hate, then they jerk away from it, as a finger jerks back from hot metal; in addition, both ganders constantly perform displacement activities, preening their feathers, shaking imaginary objects from their beaks, and so on. They cannot simply walk away, for any action suggestive of flight is forbidden by the age-old commandment to "save face" at any cost. One cannot help pitying them in their awkward situation.

The investigator of the problems of intra-specific aggression would give much to be able to determine, by an exact quantitative motivation analysis, the proportions of original aggression and autonomous triumph-cry drive contained in individual cases of this ceremony. We believe that we are gradually nearing the solution to this problem, but a description of the researches involved would take us too far here.

Let us recapitulate what has been said in the previous chapters about aggression and those special inhibiting mechanisms which, in certain permanently united individuals, not only exclude the possibility of their fighting each other but also create between them a bond such as the triumph ceremony of geese. Let us examine the relations between this bond and those of other mechanisms of community life described in the preceding chapters. As I read through these chapters for the purpose of making this summary, I realize how little I have succeeded in doing justice to the greatness and importance of the phylogenetic phenomena whose workings I think I really understand myself, but which are so difficult to explain, and I am overcome by the discouraging feeling of helplessness. One might think that a scholar with a certain gift for expressing himself, having dedicated his whole life to a specific subject, would be able to describe and communicate the results of his labors in such a way that his reader would understand not only what he knows but also what he feels about them. I can only hope that the following summary of my facts will convey to the reader at least a pale reflection of what I cannot put into words.

As we know from Chapter Eight, there are animals totally devoid of aggression which keep together for life in firmly united flocks. One would think that such animals would be predestined to develop permanent friendships and brotherly unions of individuals, and yet these characteristics are never found among such peaceable herd creatures; their association is always entirely anonymous. A personal bond, an individual friendship, is found only in animals with highly developed intra-specific aggression; in fact, this bond is the firmer, the more aggressive the particular animal and species is. There are few more aggressive fish than cichlids, few more aggressive birds than geese. Proverbially the most aggressive of all mammals, Dante's *bestia senza pace,* the wolf, is the most faithful of friends. Some animals are alternately territorial and aggressive, and nonaggressive and social, according to the season, and in these species every personal bond is limited to the period of aggressiveness. Undoubtedly the personal bond developed at that phase of evolution when, in aggressive animals, the co-operation of two or more individuals was necessary for a species-preserving purpose, usually brood tending. Doubtless the personal bond, love, arose in many cases from intra-specific aggression, by way of ritualization of a redirected attack or threatening. Since these rites are tied up with the person of the partner, and since they later become a need as independent instinct actions, they make the presence of a partner an absolute necessity and make the partner itself the "animal with home valency"—having the same emotional value as the home.

Intra-specific aggression is millions of years older than personal friendship and love. During long epochs of the earth's history, there have been animals that were certainly extraordinarily fierce and aggressive. Nearly all reptiles of the present day are aggressive and it is unlikely that those of antiquity were less so. But the personal bond is known only in certain teleost fishes, birds, and mammals, that is in groups that did not appear before the Tertiary period. Thus intra-specific aggression can certainly exist without its counterpart, love, but conversely there is no love without aggression.

A behavior mechanism that must be sharply differentiated from intra-specific aggression is hate, the ugly little brother of love. As opposed to ordinary aggression, it is directed toward one individual, just as love is, and probably hate presupposes the presence of love: one can really hate only where one has loved and, even if one denies it, still does.

It is superfluous to point out the analogies between the social behavior patterns of many animals, particularly wild geese, and those of man. All the truisms in our proverbs seem to apply equally to geese. As good evolutionists and Darwinians, we can and must draw important conclusions from this fact. We know that the youngest common ancestors of birds and mammals were very low reptiles of the Upper Devonian and Lower Carboniferous strata, which certainly had no highly developed social life and were scarcely more intelligent than frogs. Thus the similarities in the social behavior patterns of the Greylag Goose and of man are not derived from a common ancestor but have arisen by so-called convergent adaptation. They do not owe their existence to chance; this would be an improbability which could be calculated, but could only be expressed in astronomical figures.

If, in the Greylag Goose and in man, highly complex norms of behavior, such as falling in love, strife for ranking order, jealousy, grieving, etc., are not only similar but down to the most absurd details the same, we can be sure that every one of these instincts has a very special survival value, in each case almost or quite the same in the Greylag and in man. Only in this way can the conformity of behavior have developed.

As natural scientists who do not believe in "infallible instincts" or other miracles, we of course assume that every one of these behavior patterns is the function of a corresponding special physical organization of the nervous system, sense organs, etc., in other words of a structure evolved in the organism by selection pressure. If we imagine how complicated a physiological apparatus such as an electronic brain would have to be to produce a social behavior pattern like the triumph ceremony, we realize with astonishment that a wonderful organ such as the eye or the ear seems simple in comparison. The more complex and differentiated two analogously constructed and similarly functioning organs are, the more right we have to group them in the same functional conception and to call them by the same name, however different their phylogenetic origin may be. When Cephalopods, like the Octopus, Squid, and Cuttlefish, on the one hand and vertebrates on the other have invented, independently of one another, eyes built on the same principle as the lens camera, and when in both cases these organs have similar constructional units such as lens, iris, vitreous humor and retina, no reasonable person will object to calling both the organ of the Cephalopods and that of the vertebrate an eye—without any quotation marks. We are equally justified in omitting the quotation marks when speaking of the social behavior patterns of higher animals which are analogous with those of man.

All that I have said in this chapter should be a warning to the spiritual pride of many people. In an animal not even belonging to the favored class of mammals we find a behavior mechanism that keeps certain individuals together for life, and this behavior pattern has become the strongest motive governing all action; it can overcome all "animal" drives, such as hunger, sexuality, aggression, and fear, and it determines social order in its species-characteristic form. In all these points this bond is analogous with those human functions that go hand in hand with the emotions of love and friendship in their purest and noblest form.

Chapter Twelve

On the Virtue of Humility

I may claim that the contents of the preceding chapters are natural science: the recorded facts are verified, as far as it is possible to say this of the results of a science as young as comparative ethology. Now, however, we leave the record of facts evidenced by observations and experiments on the aggressive behavior of animals and turn to the question of whether they can teach us something applicable to man and useful in circumventing the dangers arising from his aggressive drives.

There are people who see in this question an insult to human dignity. All too willingly man sees himself as the center of the universe, as something not belonging to the rest of nature but standing apart as a different and higher being. Many people cling to this error and remain deaf to the wisest command ever given by a sage, the famous "Know thyself" spoken by Chilon but generally attributed to Socrates. What keeps people from listening to it? There are three obstacles, all of them motivated by strong emotions. The first is easily overcome by the man of insight; the second is at least honorable, in spite of its harmful effects; the third is understandable from the standpoint of cultural history and is therefore forgivable, but it is the most difficult to remove. All three are inseparably bound up and shot through with a most dangerous human quality, of which the proverb says that it goes before a fall: pride. I will now discuss these obstacles and try to show in what manner they are harmful, and then I will do my best to contribute toward their elimination.

The first obstacle is the most primitive. It hinders self-knowledge by inhibiting man's awareness of his own evolutionary origin. Its irrational quality and its stubborn tenacity are paradoxically derived from the great likeness which our nearest animal relations bear to us. If people did not know the chimpanzee, they would be more easily convinced of their own origin. An inexorable law of perception prevents us from seeing in the ape, particularly in the chimpanzee, an animal like other animals, and makes us see in its face the human physiognomy. From this point of view, measured by human standards, the chimpanzee of course appears as something horrible, a diabolical caricature of ourselves. In looking at the gorilla or the orangutan, which are less closely related to us, our judgment is correspondingly less distorted. The heads of the old males may look to us as bizarre devil's masks, impressive and even aesthetically appealing. This is not the way, however, that we can feel about the chimpanzee: he is irresistibly funny and at the same time as common, as vulgar as no other animal but a debased human being can ever be. This subjective impression is not altogether wrong: there are reasons for supposing that the common ancestor of man and the chimpanzee stood not lower but considerably higher than the chimpanzee does today. Absurd though the contemptuous attitude of man to the chimpanzee may be in itself, its strong emotional content has nevertheless misled several scientists into building up entirely unfounded theories about the origin of man: his evolution from animals is not disputed, but his close relationship to the repulsive chimpanzee is either passed over in a few logical skips or circumvented by sophistic detours.

The chimpanzee, however, is irresistibly funny just because he is so similar to us. What is worse is that in the narrow confinement of zoological gardens, adult chimpanzees degenerate much in the same way as human beings would under comparable circumstances and give an impression of real dissoluteness and depravity. Even the normal chimp observed in perfect health gives the impression not of an extremely highly evolved animal but rather of a degenerate and debased human being.

The second obstacle to self-knowledge is our reluctance to accept the fact that our own behavior obeys the laws of natural causation. Bernhard Hassenstein has called this attitude the "anticausal value judgment." The reluctance of many people to recognize the causal determination of all natural phenomena, human behavior included, undoubtedly stems from the justifiable wish to possess a free will and to feel that our actions are determined not by fortuitous causes but by higher aims.

A third great obstacle to human self-knowledge is—at least in our Western cultures—a heritage of idealistic philosophy. It stems from the dichotomy of the world into the external world of things, which to idealistic thought is devoid of values, and the inner world of human thought and reason to which alone values are attributed. This division appeals to man's spiritual pride. It supports him in his reluctance to accept the determination of his own behavior by natural laws. How deeply it has penetrated into accepted ways of thinking can be seen from the alteration in meaning of the words "idealist" and "realist," which originally signified philosophic attitudes but today imply moral value judgments. We must realize how common it has become in Western, particularly German, thought to consider that whatever can be explained by the laws of nature is automatically devoid of higher values. To anybody thinking in this way, explanation means devaluation.

I must here guard against the possible reproach that I am

preaching against the three obstacles to human self-knowledge because they contradict my own scientific and philosophic views: I am not arguing against the rejection of the doctrine of evolution only because I am a convinced Darwinian; my opposition to the belief that natural explanation depreciates whatever it explains is not motivated by the fact that I happen to be professionally engaged in causal analysis; nor do I object to certain consequences of idealistic thought because my own epistemological attitude is that of hypothetical realism. I have better reasons. Science is often accused of having brought terrible dangers upon man by giving him too much power over nature. This accusation would be justified only if scientists were guilty of having neglected man himself as a subject for research. The danger to modern man arises not so much from his power of mastering natural phenomena as from his powerlessness to control sensibly what is happening today in his own society. I maintain that this powerlessness is entirely the consequence of the lack of human insight into the causation of human behavior. What I intend to show is that the insight necessary to control our own social behavior is blocked by the three pride-inspired obstacles to self-knowledge.

These obstacles prevent the causal analysis of all those processes in the life of man which he regards as being of particular value, in other words those processes of which he is proud. It cannot be stressed enough: the fact that the functions of our digestive system are well known and that, owing to this knowledge, medicine, particularly intestinal surgery, saves many thousands of human lives annually, is entirely due to the fortunate circumstance that the functions of these organs do not evoke particular awe or reverence. If, on the other hand, we are powerless against the pathological disintegration of our social structure, and if, armed with atomic weapons, we cannot control our aggressive behavior any more sensibly than any animal species, this deplorable situation is largely due to our arrogant refusal to regard our own behavior as subject to the laws of nature and as accessible to causal analysis.

Science is not to blame for men's lack of self-knowledge. Giordano Bruno went to the stake because he told his fellow men that they and their planet were only a speck of dust in a cloud of countless other specks. When Charles Darwin discovered that men are descended from animals, they would have been glad to kill him and there was certainly no lack of attempts to silence him. When Sigmund Freud attempted to analyze the motives of human social behavior and to explain its causes from the subjective-psychological side, but with the method of approach of true natural science, he was accused of irreverence, blind materialism, and even pornographic tendencies. Humanity defends its own self-esteem with all its might, and it is certainly time to preach humility and to try seriously to break down all obstructions to human self-knowledge.

I will begin by attacking the resistance to the discoveries of Charles Darwin, and it may be considered an encouraging sign for the gradual spread of scientific education that today I have no longer to combat people who rise up against the findings of Giordano Bruno. I think I know a simple method of reconciling people to the fact that they are part of nature and have themselves originated by natural evolution without any infringement of natural laws: one need only show them the beauty and greatness of the universe, the awe-inspiring laws that govern it. Surely, nobody who knows enough about the phylogenetic evolution of the world of organisms can feel any inner resistance to the knowledge that he himself owes his existence to this greatest of all natural phenomena. I will not here discuss the probability, or rather the certainty, of evolution, a certainty which by far surpasses that of all our historical knowledge. Everything we know confirms the fact of evolution; it possesses, as well, everything that makes a myth of creation valuable: utter convincingness, entrancing beauty, and awe-inspiring greatness.

Anyone who understands this cannot possibly be repulsed by Darwin's recognition of the fact that we have a common origin with animals, or by Freud's realization that we are still driven by the same instincts as our prehuman ancestors. Quite the contrary: this knowledge inspires a new feeling of respect for the functions of reason and moral responsibility which first came into the world with man and which, provided he does not blindly and arrogantly deny the existence of his animal inheritance, give him the power to control it.

A further reason why some people still resist the doctrine of evolution is the great respect we human beings have for our ancestors. To descend from, means, literally, to come down, and even in Roman law it was customary to put the ancestor uppermost in the pedigree and to draw the family tree branching downward. The fact that a human being has only two parents but 256 great-great-great-great-great-great-grandparents does not appear in such pedigrees even if they extend to many generations. We avoid mentioning this multitude because among so many ancestors we would not find enough of whom we could be proud. According to some authors, the term "descent" may derive from the fact that in ancient times man was fond of tracing his origin to the gods. That the family tree of life grows not from above downward but from below upward escaped man's notice until Darwin's time; thus the word "descent" stands for the opposite of what it means, unless we wish to take it literally that our forefathers, in their time, came down from the trees. This they actually did, though, as we know today, a long time before they became human beings.

The terms "development" and "evolution" are nearly as inadequate as "descent." They too came into use at a time when we knew nothing of the creative processes of the origin of species and only knew about the origin of individuals from eggs or seeds. A chick literally develops from an egg and a sunflower from a seed; that is, nothing originates from the germ that was not already formed inside it.

The growth of the great family tree of life is quite different. Though the ancestral form is the indispensable prerequisite for the origin of its more highly developed descendants, their evolution can in no way be predicted from the characters of the ancestor. The fact that birds evolved from reptiles or man from apes is a historically unique achievement of evolution. By laws that govern every living being, evolution has a general trend to the higher but, in all its details, is determined by so-called chance, that is by innumerable collateral chains of causation which in principle can never be completely apprehended. It is by "chance" in this sense that from primitive forebears in Australia, eucalyptus trees and kangaroos originated, and in Europe and Asia, oak trees and man. The newly evolved form of life is an achievement, and its characters cannot be predicted from those of its forebear; that is, in the large majority of cases, something higher than the latter. The naïve value judgment expressed in the words "lower animals" is for every unbiased person an inevitable necessity of thinking and feeling.

The scientist who considers himself absolutely "objective" and believes that he can free himself from the compulsion of the "merely" subjective should try—only in imagination, of course—to kill in succession a lettuce, a fly, a frog, a guinea pig, a cat, a dog, and finally a chimpanzee. He will then be aware how increasingly difficult murder becomes as the victim's level or organization rises. The degree of inhibition against killing each one of these beings is a very precise measure for the considerably different values that we cannot help attributing to lower and higher forms of life. To any man who finds it equally easy to chop up a live dog and a live lettuce I would recommend suicide at his earliest convenience!

The principle that science should be indifferent to values must not lead to the belief that evolution, that most wonderful of all chains of naturally explicable processes, is not capable of creating new values. That the origin of a higher form of life from a simpler ancestor means an increase in values is a reality as undeniable as that of our own existence.

None of our Western languages has an intransitive verb to do justice to the increase of values produced by very nearly every step in evolution. One cannot possibly call it development when something new and higher arises from an earlier stage which does not contain the constituent properties for the new and higher being. Fundamentally this applies to every bigger step in the genesis of the world of organisms, including the first step, the origin of life, and the most recent one—the origin of man from the anthropoid.

In spite of all epoch-making and inspiring new discoveries in biochemistry and virology, the origin of life is still the most puzzling of all natural phenomena. The difference between the processes of life and those occurring in nonliving matter can only be defined by what B. Hassenstein has termed an "injunctive" definition. This means that to define the concept of life it is necessary to enumerate a number of constituent characteristics, none of which, taken by itself, constitutes life but which, taken all together, in their summation and interaction, do indeed represent the essence of life. For each of the processes of life, such as metabolism, growth, propagation, and so on, an analogy can be found in inorganic matter, but all together can only be found in the living protoplasm. We are thus justified in maintaining that life processes are chemical and physical processes, and as such there is no doubt that fundamentally there is a natural explanation for them. No miracle is required to explain their peculiarities, for these are adequately explained by the complicated nature of molecular and other structures.

It is wrong, however, to assert that life processes are essentially chemical and physical processes. This assertion, though often made, contains unnoticed a false value judgment. The very "essence" of life processes is that combination of characteristics which constitute their "*injunctive*" definition, and it is with regard to these very characteristics that life processes are emphatically not what we ordinarily mean when we speak of chemical or physical processes. By virtue of the molecular structure of the living matter in which they take place, the processes of life fulfill a great number of very particular functions, such as self-regulation, self-preservation, acquisition and storage of information, and, above all, reproduction of the structures essential for these functions. These, though in principle causally explicable, cannot take place in other, structurally less complex matter.

In the world of organisms, the relation of every higher life form to the lower one from which it originated is fundamentally the same as the relation of the processes and structures of life to those of the nonliving. It would be a gross misrepresentation to say that the bird's wing is "nothing but" a reptilian forelimb, or, still worse, to say that man is "nothing but" an ape. Indeed he is one, but he is much more besides: he is *essentially* more.

A sentimental misanthropist coined the often cited aphorism "The more I see of human beings, the more I like animals." I maintain the contrary: only the person who knows animals, including the highest and most nearly related to ourselves, and who has gained insight into evolution, will be able to apprehend the unique position of man. We are the highest achievement reached so far by the great constructors of evolution. We are their "latest" but certainly not their last word. The scientist must not regard anything as absolute, not even the laws of pure reason. He must remain aware of the great fact, discovered by Heraclitus, that nothing whatever really remains the same even for one moment but that everything is perpetually changing. To regard man, the most ephemeral and rapidly evolving of all species, as the final and unsurpassable achievement of creation, especially at his present-day particularly dangerous and disagreeable stage of development, is certainly the most arrogant and dangerous of all untenable doctrines. If I thought of man as the final image of God, I should not know what to think of God. But when I consider that our ancestors, at a time fairly recent in relation to the earth's history, were perfectly ordinary apes, closely related to chimpanzees, I see a glimmer of hope. It does not require considerable optimism to assume that from us human beings something better and higher may evolve. Far from seeing in man the irrevocable and unsurpassable image of God, I assert—more modestly and, I believe, in greater awe of the Creation and its infinite possibilities—that the long-sought missing link between animals and the really humane being is ourselves!

The first great obstacle to human self-knowledge, the reluctance to believe in our evolution from animals, is based, as I have tried to show, on ignorance or misunderstanding of the essence of organic creation. Fundamentally at least, it should be possible to remove this obstacle by teaching and learning. Similar means should help to remove the second obstacle, now to be discussed, the antipathy toward the causal determination; but in this case the misunderstanding is far more difficult to clear up. Its root is the basically erroneous belief that a process which is causally determined cannot at the same time be goal-directed. Admittedly, there are countless processes in the universe which are not goal-directed, and in these cases the question "What for?" must remain unanswered, unless we are determined to find an answer at any price, in measureless overestimation of the importance of man, as, for instance, if we explain the rising of the moon as a switching on of night illumination for our especial benefit. There is, however, no process to which the question of causes cannot be applied.

As already stated in Chapter Three, the question "What is it for?" makes sense only where the great constructors—or a living constructor constructed by them—have been at work. Only where parts of a systemic whole have become specialized, by division of labor, for different functions, each completing the other, does the question "What is it for?" make any sense. This holds good for life processes, as also for those lifeless structures and functions which the living being makes use of for its own purposes, for instance, man-made machines. In these cases the question "What is it for?" is not only relevant but absolutely necessary. We could not understand the cause of the cat's sharp claws if we had not first found out that their special function was catching mice.

At the beginning of Chapter Six, on the great parliament of instincts, we have already said that the answering of the question "What is it for?" does not rule out the question of the cause. How little the two questions preclude each other can be shown by a simple analogy. I am driving through the countryside in my old car, to give a lecture in a distant town, and I ponder on the usefulness of my car, the goals or aims which are so well served by its construction, and it pleases me to think how all this contributes to achieve the purpose of my journey. Suddenly the motor coughs once or twice and peters out. At this stage I am painfully aware that the reason for my journey does not make my car go; I am learning the hard way that aims or goals are not causes. It will now be well for me to concentrate exclusively on the natural causes of the car's workings, and to find out at what stage the chain of their causation was so unpleasantly interrupted.

Medicine, the "queen of applied sciences," furnishes us even better examples of the erroneousness of the view that purposiveness and causality preclude each other. No "life purpose," no "whole-making factor," and no sense of imperative obligation can help the unfortunate patient with acute appendicitis, but even the youngest hospital surgeon can help him if he has rightly diagnosed the cause of the trouble. The appreciation of the fact that life processes are directed at aims or goals, and the realization of the other fact that they are, at the same time, determined by causality, not only do not preclude each other but they only make sense in combination. If man did not strive toward goals, his questions as to causes would have no sense; if he has no insight into cause and effect, he is powerless to guide effects toward determined goals, however rightly he may have understood the meaning of these goals.

This relation between the purposive and the causal aspects of life processes seems to me quite obvious, but evidently many people are under the illusion of their incompatibility. A classic example of how even a great mind can be a victim of this illusion is seen in the works of W. McDougall, the founder of "purposive psychology." In his book, *Outline of Psychology,* he rejects every causal physiological explanation of animal behavior, with one exception: he explains the misfunctioning of the light-compass-orientation, which causes insects to fly at night into flames, by so-called tropisms, that is causally analyzed orientation mechanisms.

Probably the reason why people are so afraid of causal considerations is that they are terrified lest insight into the causes of earthly phenomena should expose man's free will as an illusion. In reality the fact that I have a will is as undeniable as the fact of my existence. Deeper insight into the physiological concatenation of causes of my own behavior cannot in the least alter the fact *that* I will but it can alter *what* I will.

Only on very superficial consideration does free will seem to imply that "we can want what we will" in complete lawlessness, though this thought may appeal to those who flee as in claustrophobia from causality. We must remember how the theory of indeterminism of microphysical phenomena, the "acausal" quantum physics, was avidly seized and on its foundations hypotheses built up to mediate between physical determinism and belief in free will, though the only freedom thereby left to the will was the lamentable liberty of the fortuitously cast die. Nobody can seriously believe that free will means that it is left entirely to the will of the individual, as to an irresponsible tyrant, to do or not do whatever he pleases. Our freest will underlies strict moral laws, and one of the reasons for our longing for freedom is to prevent our obeying other laws than these. It is significant that the anguished feeling of not being free is never evoked by the realization that our behavior is just as firmly bound to moral laws as physiological processes are to physical ones. We are all agreed that the greatest and most precious freedom of man is identical with the moral laws within him. Increasing knowledge of the natural causes of his own behavior can certainly increase a man's faculties and enable him to put his free will into action, but it can never diminish his will. If, in the impossible case of a utopian, complete, and ultimate success of causal analysis, man ever should achieve complete insight into the causality of earthly phenomena, including the workings of his own organism, he would not cease to have a will but it would be in perfect harmony with the incontrovertible lawfulness of the universe, the *Weltvernunft* of the Logos. This idea is foreign only to our present-day Western thought; it was quite familiar to ancient Indian philosophy and to the mystics of the Middle Ages.

I now come to the third great obstacle to human self-knowledge, to the belief—deeply rooted in our Western culture—that what can be explained in terms of natural science has no values. This belief springs from an exaggeration of Kant's value philosophy, the consequence of the idealistic dichotomy of the world into the external world of things and the internal laws of human reason. As already intimated, fear of causality is one of the emotional reasons for the high values set on the unfathomable, but other unconscious factors are also involved. The behavior of the ruler, the father figure, whose essential features include an element of arbitrariness and injustice, is unaccountable. God's decree is inscrutable. Whatever can be explained by natural causes can be controlled, and with its obscurity it loses most of its terror. Benjamin Franklin made of the thunderbolt, the instrument of Zeus's unaccountable whim, an electric spark against which the lightning conductors of our houses can protect us. The unfounded fear that nature might be desecrated by causal insight forms the second chief motive of people's fear of causality. Hence there arises a further obstacle to science, and this is all the stronger the greater a man's sense of the aesthetic beauty and awe-inspiring greatness of the universe and the more beautiful and venerable any particular natural phenomenon seems to him.

The obstacle to research arising from these unfortunate associations is the more dangerous since it never crosses the threshold of consciousness. If questioned, such people would profess in all sincerity to be supporters of scientific research, and, within the limits of a circumscribed special field, they may even be great scientists. But subconsciously they are firmly resolved not to carry their natural explanations beyond the limits of the awe-inspiring. Their error does not lie in the false assumption that some things are inexplorable: nobody knows so well as the scientist that there are limits to human understanding, but he is always aware that we do not know where these limits lie. Kant says, "Our observation and analysis of its phenomena penetrate to the depth of nature. We do not know how far this will lead us in time." The obstacle to scientific research produced by this line of thought consists in setting a dogmatic border between the explorable and what is considered beyond exploration. Many excellent observers have so great a respect for life and its characteristics that they draw the line at its origin. They accept a special life force, *force vitale,* a direction-giving, whole-making factor which, they consider, neither requires nor permits an explanation. Others draw the line where they feel that human dignity demands a halt before any further attempts at natural explanation.

The attitude of the true scientist toward the real limits of human understanding was unforgettably impressed on me in early youth by the obviously unpremeditated words of a great biologist: Alfred Kühn finished a lecture to the Austrian Academy of Sciences with Goethe's words, "It is the greatest joy of the man of thought to have explored the explorable and then calmly to revere the inexplorable." After the last word he hesitated, raised his hand in repudiation, and cried, above the applause, "No, *not* calmly, gentlemen; not *calmly!*" One could even define a true scientist by his ability to feel undiminished awe for the explorable that he has explored; from this arises his ability to want to explore the apparently inexplorable: he is not afraid of desecrating nature by causal insight. Never has natural explanation of one of its marvelous processes exposed nature as a charlatan who has lost the reputation of his sorcery; natural causal associations have always turned out to be grander and more awe-inspiring than even the most imaginative mythical interpretation. The true scientist does not need the inexplorable, the supernatural, to evoke his reverence: for him there is only one miracle, namely, that everything, even the finest flowerings of life, have come into being without miracles; for him the universe would lose some of its grandeur if he thought that any phenomenon, even reason and moral sense in noble-minded human beings, could be accounted for only by an *infringement* of the omnipresent and omnipotent laws of *one* universe.

Nothing can better express the feelings of the scientist toward the great unity of the laws of nature than Immanuel

Kant's words: "Two things fill the mind with ever new and increasing awe: the stars above me and the moral law within me." Admiration and awe did not prevent the great philosopher from finding a natural explanation for the laws of the heavens, indeed an explanation based on their evolutionary origin. Would he, who did not yet know of the evolution of the world of organisms, be shocked that we consider the moral law within us not as something given, a priori, but as something which has arisen by natural evolution, just like the laws of the heavens?

Chapter Thirteen

Ecce Homo!

Let us imagine that an absolutely unbiased investigator on another planet, perhaps on Mars, is examining human behavior on earth, with the aid of a telescope whose magnification is too small to enable him to discern individuals and follow their separate behavior, but large enough for him to observe occurrences such as migrations of peoples, wars, and similar great historical events. He would never gain the impression that human behavior was dictated by intelligence, still less by responsible morality. If we suppose our extraneous observer to be a being of pure reason, devoid of instincts himself and unaware of the way in which all instincts in general and aggression in particular can miscarry, he would be at a complete loss how to explain history at all. The ever-recurrent phenomena of history do not have reasonable causes. It is a mere commonplace to say that they are caused by what common parlance so aptly terms "human nature." Unreasoning and unreasonable human nature causes two nations to compete, though no economic necessity compels them to do so; it induces two political parties or religions with amazingly similar programs of salvation to fight each other bitterly, and it impels an Alexander or a Napoleon to sacrifice millions of lives in his attempt to unite the world under his scepter. We have been taught to regard some of the persons who have committed these and similar absurdities with respect, even as "great" men, we are wont to yield to the political wisdom of those in charge, and we are all so accustomed to these phenomena that most of us fail to realize how abjectly stupid and undesirable the historical mass behavior of humanity actually is.

Having realized this, however, we cannot escape the question why reasonable beings do behave so unreasonably. Undeniably, there must be superlatively strong factors which are able to overcome the commands of individual reason so completely and which are so obviously impervious to experience and learning. As Hegel said, "What experience and history teach us is this—that people and governments never have learned anything from history, or acted on principles deduced from it."

All these amazing paradoxes, however, find an unconstrained explanation, falling into place like the pieces of a jigsaw puzzle, if one assumes that human behavior, and particularly human social behavior, far from being determined by reason and cultural tradition alone, is still subject to all the laws prevailing in all phylogenetically adapted instinctive behavior. Of these laws we possess a fair amount of knowledge from studying the instincts of animals. Indeed, if our extramundane observer were a knowledgeable ethologist, he would unavoidably draw the conclusion that man's social organization is very similar to that of rats, which, like humans, are social and peaceful beings within their clans, but veritable devils toward all fellow members of their species not belonging to their own community. If, furthermore, our Martian naturalist knew of the explosive rise in human populations, the ever-increasing destructiveness of weapons, and the division of mankind into a few political camps, he would not expect the future of humanity to be more rosy than that of several hostile clans of rats on a ship almost devoid of food. And this prognosis would even be optimistic, for in the case of rats, reproduction stops automatically when a certain state of overcrowding is reached while man as yet has no workable system for preventing the so-called population explosion. Furthermore, in the case of the rats it is likely that after the wholesale slaughter enough individuals would be left over to propagate the species. In the case of man, this would not be so certain after the use of the hydrogen bomb.

It is a curious paradox that the greatest gifts of man, the unique faculties of conceptual thought and verbal speech which have raised him to a level high above all other creatures and given him mastery over the globe, are not altogether blessings, or at least are blessings that have to be paid for very dearly indeed. All the great dangers threatening humanity with extinction are direct consequences of conceptual thought and verbal speech. They drove man out of the paradise in which he could follow his instincts with impunity and do or not do whatever he pleased. There is much truth in the parable of the tree of knowledge and its fruit, though I want to make an addition to it to make it fit into my own picture of Adam: that apple was thoroughly unripe! Knowledge springing from conceptual thought robbed man of the security provided by his well-adapted instincts long, long before it was sufficient to provide him with an equally safe adaptation. Man is, as Arnold Gehlen has so truly said, by nature a jeopardized creature.

Conceptual thought and speech changed all man's evolution by achieving something which is equivalent to the inheritance of acquired characters. We have forgotten that the verb "inherit" had a juridic connotation long before it acquired a biological one. When a man invents, let us say, bow and arrow, not only his progeny but his entire community will inherit the knowledge and the use of these tools and possess them just as surely as organs grown on the body. Nor is their loss any more likely than the rudimentation of an organ of equal survival value. Thus, within one or two generations a process of ecological adaptation can be achieved which, in normal phylogeny and without the interference of conceptual thought, would have taken a time of an altogether different, much greater order of magnitude. Small wonder, indeed, if the evolution of social instincts and, what is even more important, social inhibitions could not keep pace with the rapid development forced on human society by the growth of traditional culture, particularly material culture.

Obviously, instinctive behavior mechanisms failed to cope with the new circumstances which culture unavoidably produced even at its very dawn. There is evidence that the first inventors of pebble tools, the African Australopithecines, promptly used their new weapon to kill not only game, but fellow members of their species as well. Peking Man, the Prometheus who learned to preserve fire, used it to roast his brothers: beside the first traces of the regular use of fire lie the mutilated and roasted bones of Sinanthropus pekinensis himself.

One is tempted to believe that every gift bestowed on man by his power of conceptual thought has to be paid for with a dangerous evil as the direct consequence of it. Fortunately for us, this is not so. Besides the faculty of conceptual thought, another constituent characteristic of man played an important role in gaining a deeper understanding of his environment, and this is curiosity. Insatiable curiosity is the root of exploration and experimentation, and these activities, even in their most primitive form, imply a function akin to asking questions. Explorative experimentation is a sort of dialogue with surrounding nature. Asking a question and recording the answer leads to anticipating the latter, and, given conceptual thought, to the linking of cause and effect. From hence it is but a step to consciously foreseeing the consequences of one's actions. Thus, the same human faculties which supplied man with tools and with power dangerous to himself, also gave him the means to prevent their misuse: rational responsibility. I shall now proceed to discuss, one by one, the dangers which humanity incurs by rising above the other animals by virtue of its great, specific gifts. Subsequently I shall try to show in what way the greatest gift of all, rational, responsible morality, functions in banning these dangers. Most important of all, I shall have to expound the functional limitation of morality.

In the chapter on behavior mechanisms functionally analogous to morality, I have spoken of the inhibitions controlling aggression in various social animals, preventing it from injuring or killing fellow members of the species. As I explained, these inhibitions are most important and consequently most highly differentiated in those animals which are capable of killing living creatures of about their own size. A raven can peck out the eye of another with one thrust of its beak, a wolf can rip the jugular vein of another with a single bite. There would be no more ravens and no more wolves if reliable inhibitions did not prevent such actions. Neither a dove nor a hare nor even a chimpanzee is able to kill its own kind with a single peck or bite; in addition, animals with relatively poor defense weapons have a correspondingly great ability to escape quickly, even from specially armed predators which are more efficient in chasing, catching, and killing than even the strongest of their own species. Since there rarely is, in nature, the possibility of such an animal's seriously injuring one of its own kind, there is no selection pressure at work here to breed in killing inhibitions. The absence of such inhibitions is apparent to the animal keeper, to his own and to his animals' disadvantage, if he does not take seriously the intra-specific fights of completely "harmless" animals. Under the unnatural conditions of captivity, where a defeated animal cannot escape from its victor, it may be killed slowly and cruelly. In my book *King*

Solomon's Ring, I have described in the chapter "Morals and Weapons" how the symbol of peace, the dove, can torture one of its own kind to death, without the arousal of any inhibition.

Anthropologists concerned with the habits of Australopithecus have repeatedly stressed that these hunting progenitors of man have left humanity with the dangerous heritage of what they term "carnivorous mentality." This statement confuses the concepts of the carnivore and the cannibal, which are, to a large extent, mutually exclusive. One can only deplore the fact that man has definitely not got a carnivorous mentality! All his trouble arises from his being a basically harmless, omnivorous creature, lacking in natural weapons with which to kill big prey, and, therefore, also devoid of the built-in safety devices which prevent "professional" carnivores from abusing their killing power to destroy fellow members of their own species. A lion or a wolf may, on extremely rare occasions, kill another by one angry stroke, but, as I have already explained in the chapter on behavior mechanisms functionally analogous to morality, all heavily armed carnivores possess sufficiently reliable inhibitions which prevent the self-destruction of the species.

In human evolution, no inhibitory mechanisms preventing sudden manslaughter were necessary, because quick killing was impossible anyhow; the potential victim had plenty of opportunity to elicit the pity of the aggressor by submissive gestures and appeasing attitudes. No selection pressure arose in the prehistory of mankind to breed inhibitory mechanisms preventing the killing of conspecifics until, all of a sudden, the invention of artificial weapons upset the equilibrium of killing potential and social inhibitions. When it did, man's position was very nearly that of a dove which, by some unnatural trick of nature, has suddenly acquired the beak of a raven. One shudders at the thought of a creature as irascible as all prehuman primates are, swinging a well-sharpened hand-ax. Humanity would indeed have destroyed itself by its first inventions, were it not for the very wonderful fact that inventions and responsibility are both the achievements of the same specifically human faculty of asking questions.

Not that our prehuman ancestor, even at a stage as yet devoid of moral responsibility, was a fiend incarnate; he was by no means poorer in social instincts and inhibitions than a chimpanzee, which, after all, is—his irascibility not withstanding—a social and friendly creature. But whatever his innate norms of social behavior may have been, they were bound to be thrown out of gear by the invention of weapons. If humanity survived, as, after all, it did, it never achieved security from the danger of self-destruction. If moral responsibility and unwillingness to kill have indubitably increased, the ease and emotional impunity of killing have increased at the same rate. The distance at which all shooting weapons take effect screens the killer against the stimulus situation which would otherwise activate his killing inhibitions. The deep, emotional layers of our personality simply do not register the fact that the crooking of the forefinger to release a shot tears the entrails of another man. No sane man would even go rabbit hunting for pleasure if the necessity of killing his prey with his natural weapons brought home to him the full, emotional realization of what he is actually doing.

The same principle applies, to an even greater degree, to the use of modern remote-control weapons. The man who presses the releasing button is so completely screened against seeing, hearing, or otherwise emotionally realizing the consequences of his action, that he can commit it with impunity —even if he is burdened with the power of imagination. Only thus can it be explained that perfectly good-natured men, who would not even smack a naughty child, proved to be perfectly able to release rockets or to lay carpets of incendiary bombs on sleeping cities, thereby committing hundreds and thousands of children to a horrible death in the flames. The fact that it is good, normal men who did this, is as eerie as any fiendish atrocity of war!

As an indirect consequence, the invention of artificial weapons has brought about a most undesirable predominance of intra-specific selection within mankind. In the third chapter, in which I discussed the survival value of aggression, and also in the tenth, dealing with the structure of society in rats, I have already spoken of the manner in which competition between the fellow members of one species can produce unadaptive results when it exerts a selection pressure totally unrelated to extra-specific environment (Chapter Three, pages 41 ff).

When man, by virtue of his weapons and other tools, of his clothing and of fire, had more or less mastered the inimical forces of his extra-specific environment, a state of affairs must have prevailed in which the counter-pressures of the hostile neighboring hordes had become the chief selecting factor determining the next steps of human evolution. Small wonder indeed if it produced a dangerous excess of what has been termed the "warrior virtues" of man.

In 1955 I wrote a paper, "On the Killing of Members of the Same Species": "I believe—and human psychologists, particularly psychoanalysts, should test this—that present-day civilized man suffers from insufficient discharge of his aggressive drive. It is more than probable that the evil effects of the human aggressive drives, explained by Sigmund Freud as the results of a special death wish, simply derive from the fact that in prehistoric times intra-specific selection bred into man a measure of aggression drive for which in the social order of today he finds no adequate outlet." If these words contain an element of reproach against psychoanalysis, I must here withdraw them. At the time of writing, there were already some psychoanalysts who did not believe in the death wish and rightly explained the self-destroying effects of aggression as misfunctions of an instinct that was essentially life-preserving. Later, I came to know one psychiatrist and psychoanalyst who, even at that time, was examining the problem of the hypertrophy of aggression owing to intra-specific selection.

Sydney Margolin, in Denver, Colorado, made very exact psychoanalytical and psycho-sociological studies on Prairie Indians, particularly the Utes, and showed that these people suffer greatly from an excess of aggression drive which, under the ordered conditions of present-day North American Indian reservations, they are unable to discharge. It is Margolin's opinion that during the comparatively few centuries when Prairie Indians led a wild life consisting almost entirely of war and raids, there must have been an extreme selection pressure at work, breeding extreme aggressiveness. That this produced changes in the hereditary pattern in such a short time is quite possible. Domestic animals can be changed just as quickly by purposeful selection. Margolin's assumption is supported by the fact that Ute Indians now growing up under completely different educational influences suffer in exactly the same way as the older members of their tribe who grew up under the educational system of their own culture; moreover, the pathological symptoms under discussion are seen only in those Prairie Indians whose tribes were subjected to the selection process described.

Ute Indians suffer more frequently from neurosis than any other human group, and again and again Margolin found that the cause of the trouble was undischarged aggression. Many of these Indians feel and describe themselves as ill, and when asked what is the matter with them they can only say, "I am a Ute!" Violence toward people not of their tribe, and even manslaughter, belong to the order of the day, but attacks on members of the tribe are extremely rare, for they are prevented by a taboo the severity of which it is easy to understand, considering the early history of the Utes: a tribe constantly at war with neighboring Indians and, later on, with the white man, must avoid at all costs fights between its own members. Anyone killing a member of the tribe is compelled by strict tradition to commit suicide. This commandment was obeyed even by a Ute policeman who had shot a member of his tribe in self-defense while trying to arrest him. The offender, while under the influence of drink, had stabbed his father in the femoral artery, causing him to bleed to death. When the policeman was ordered by his sergeant to arrest the man for manslaughter—it was obviously not murder—he protested, saying that the man would want to die since he was bound by tradition to commit suicide and would do so by resisting arrest and forcing the policeman to shoot him. He, the policeman, would then have to commit suicide himself. The more than short-sighted sergeant stuck to his order, and the tragedy took place exactly as predicted. This and other of Margolin's records read like Greek tragedies: an inexorable fate forces crime upon people and then compels them to expiate voluntarily their involuntarily acquired guilt.

It is objectively convincing, indeed it is proof of the correctness of Margolin's interpretation of the behavior of Ute Indians, that these people are particularly susceptible to accidents. It has been proved that accident-proneness may result from repressed aggression, and in these Utes the rate of motor accidents exceeds that of any other car-driving human group. Anybody who has ever driven a fast car when really angry knows—in so far as he is capable of self-observation in this condition—what strong inclination there is to self-destructive behavior in a situation like this. Here even the expression "death wish" seems apt.

It is self-evident that intra-specific selection is still working today in an undesirable direction. There is a high positive selection premium on the instinctive foundations conducive to such traits as the amassing of property, self-assertion, etc., and there is an almost equally high negative premium on simple goodness. Commercial competition today might threaten to fix hereditarily in us hypertrophies of these traits, as horrible as the intra-specific aggression evolved by competition between warfaring tribes of Stone Age man. It is fortunate that the accumulation of riches and power does not necessarily lead to large families—rather the opposite—or else the future of mankind would look even darker than it does.

Aggressive behavior and killing inhibitions represent only one special case among many in which phylogenetically adapted behavior mechanisms are thrown out of balance by the rapid change wrought in human ecology and sociology by cultural development. In order to explain the function of responsible morality in re-establishing a tolerable equilibrium between man's instincts and the requirements of a culturally evolved social order, a few words must first be said about social instincts in general. It is a widely held opinion, shared by some contempory philosophers, that all human behavior patterns which serve the welfare of the community, as opposed to that of the individual, are dictated by specifically human rational thought. Not only is this opinion erroneous, but the very opposite is true. If it were not for a rich endowment of social instincts, man could never have risen above the animal world. All specifically human faculties, the power of speech, cultural tradition, moral responsibility, could have evolved only in a being which, before the very dawn of conceptual thinking, lived in well-organized communities. Our prehuman ancestor was indubitably as true a friend to his friend as a chimpanzee or even a dog, as tender and solicitous to the young of his community and as self-sacrificing in its defense, aeons before he developed conceptual thought and became aware of the consequences of his actions.

According to Immanuel Kant's teachings on morality, it is

human reason (*Vernunft*) alone which supplies the categorical imperative "thou shalt" as an answer to responsible self-questioning concerning any possible consequences of a certain action. However, it is doubtful whether "reason" is the correct translation of Kant's use of the word *"Vernunft,"* which also implies the connotation of common sense and of understanding and appreciation of another "reasonable" being. For Kant it is self-evident that one reasonable being cannot possibly want to hurt another. This unconscious acceptance of what he considered evident, in other words common sense, represents the chink in the great philosopher's shining armor of pure rationality, through which emotion, which always means an instinctive urge, creeps into his considerations and makes them more acceptable to the biologically minded than they would otherwise be. It is hard to believe that a man will refrain from a certain action which natural inclination urges him to perform only because he has realized that it involves a logical contradiction. To assume this, one would have to be an even more unworldly German professor and an even more ardent admirer of reason than Immanuel Kant was.

In reality, even the fullest rational insight into the consequences of an action and into the logical consistency of its premise would not result in an imperative or in a prohibition, were it not for some emotional, in other words instinctive, source of energy supplying motivation. Like power steering in a modern car, responsible morality derives the energy which it needs to control human behavior from the same primal powers which it was created to keep in rein. Man as a purely rational being, divested of his animal heritage of instincts, would certainly not be an angel—quite the opposite.

Supposing that a being entirely indifferent to values, unable to see anything worth preserving in humanity, in human culture, and in life itself, were examining the principle of its action in pressing the button releasing the hydrogen bomb and destroying all life on our planet, even a full realization of the consequences would, in such a monster, elicit no imperative forbidding the deed, but only a reaction tantamount to saying, "So what?" We need not even suppose this hypothetical creature to be actively evil and to share the view of Goethe's Mephistopheles that everything created is worthy of annihilation; mere absence of any emotional appreciation of values could make it react in the way described.

Always and everywhere it is the unreasoning, emotional appreciation of values that adds a plus or a minus sign to the answer of Kant's categorical self-questioning and makes it an imperative or a veto. By itself, reason can only devise means to achieve otherwise determined ends; it cannot set up goals or give us orders. Left to itself, reason is like a computer into which no relevant information conducive to an important answer has been fed; logically valid though all its operations may be, it is a wonderful system of wheels within wheels, without a motor to make them go round. The motive power that makes them do so stems from instinctive behavior mechanisms much older than reason and not directly accessible to rational self-observation. They are the source of love and friendship, of all warmth of feeling, of appreciation of beauty, of the urge to artistic creativeness, of insatiable curiosity striving for scientific enlightenment. These deepest strata of the human personality are, in their dynamics, not essentially different from the instincts of animals, but on their basis human culture has erected all the enormous superstructure of social norms and rites whose function is so closely analogous to that of phylogenetic ritualization. Both phylogenetically and culturally evolved norms of behavior represent motives and are felt to be values by any normal human being. Both are woven into an immensely complicated system of universal interaction to analyze which is all the more difficult as most of its processes take place in the subconscious and are by no means directly accessible to self-observation. Yet it is imperative for us to understand the dynamics of this system, because insight into the nature of values offers the only hope for our ever creating the new values and ideals which our present situation needs so badly.

Even the first compensatory function of moral responsibility, preventing the Australopithecines from destroying themselves with their first pebble tools, could not have been achieved without an instinctive appreciation of life and death. Some of the most intelligent and most social birds and mammals react in a highly dramatic way to the sudden death of a member of their species. Greylag geese will stand with outspread wings over a dying friend hissing defensively, as Heinroth saw after having shot a goose in the presence of its family. I observed the same behavior on the occasion of an Egyptian goose killing a greylag gosling by hitting it on the head with its wing; the gosling staggered toward its parents and collapsed, dying of cerebral hemorrhage. Though the parents could not have seen the deadly blow, they reacted in the described way. In the Munich zoo some years ago an essentially friendly bull elephant while playing with his keeper unintentionally injured him severely, severing an artery in the man's thigh. The elephant immediately seemed to realize that something dangerous had befallen his friend and with the best intentions did the worst thing he could do: he stood protectively over the fallen man, thus preventing medical aid from reaching him. Professor Bernhard Grzimek told me that an adult male chimpanzee, after having bitten him rather badly, seemed very concerned, after his rage had abated, about what he had done and tried to press together, with his fingers, the lips of Grzimek's worst wounds. It is highly characteristic of that dauntless scientist that he permitted the ape to do so.

It is safe to assume that the first Cain, after having stricken a fellow member of his horde with a pebble tool, was deeply concerned about the consequences of his action. He may have struck with very little malice, just as a two-year-old child may hit another with a heavy and hard object without foreseeing the effect. He may have been most painfully surprised when his friend failed to get up again; he may even have tried to help him get up, as the bull elephant is reported to have done. In any case we are safe in assuming that the first killer fully realized the enormity of his deed. There was no need for the information being slowly passed around that the horde loses dangerously in fighting potential if it slaughters too many of its members for the pot.

Whatever the consequences may have been that prevented the first killers from repeating their deed, realization of these consequences and, therewith, a primitive form of responsibility must have been at work. Apart from maintaining the equilibrium between the ability and the inhibition to kill, responsible morality does not seem to have been too severely taxed in the earliest communities of true men. It is no daring speculation to assume that the first human beings which really represented our own species, those of Cro-Magnon, had roughly the same instincts and natural inclinations as we have ourselves. Nor is it illegitimate to assume that the structure of their societies and their tribal warfare was roughly the same as can still be found in certain tribes of Papuans in central New Guinea. Every one of their tiny settlements is permanently at war with the neighboring villages; their relationship is described by Margaret Mead as one of mild reciprocal head-hunting, "mild" meaning that there are no organized raids for the purpose of removing the treasured heads of neighboring warriors, but only the occasional taking of the heads of women and children encountered in the woods.

Now let us suppose that our assumption is correct and that the men of such a paleolithic tribe did indeed have the same natural inclinations, the same endowment with social instincts as we have ourselves; let us imagine a life, lived dangerously in the exclusive company of a dozen or so close friends and their wives and children. There would be some friction, some jealousy about girls, or rank order, but on the whole I think that this kind of rivalry would come second to the continuous necessity for mutual defense against hostile neighboring tribes. The men would have fought side by side from earliest memory; they would have saved each other's lives many times; all would have ample opportunity to discharge intra-specific aggression against their enemies, none would feel the urge to injure a member of his own community. In short, the sociological situation must have been, in a great many respects, comparable to that of the soldiers of a small fighting unit on a particularly dangerous and independent assignment. We know to what heights of heroism and utter self-abnegation average, unromantic modern men have risen under these circumstances. Incidentally, it is quite typical of man that his most noble and admirable qualities are brought to the fore in situations involving the killing of other men, just as noble as they are. However cruel and savage such a community may be to another, within its bonds natural inclination alone is very nearly sufficient to make men obey the Ten Commandments—perhaps with the exception of the third. One does not steal another man's rations or weapons, and it seems rather despicable to covet the wife of a man who has saved one's life a number of times. One would certainly not kill him, and one would, from natural inclination, honor not only father and mother, but the aged and experienced in general, just as deer and baboons do, according to the observations of Fraser Darling, Washburn, and De Vore.

The imagination of man's heart is not really evil from his youth up, as we read in Genesis. Man can behave very decently indeed in tight spots, provided they are of a kind that occurred often enough in the paleolithic period to produce phylogenetically adapted social norms to deal with the situation. Loving your neighbor as yourself or risking your life in trying to save his is a matter of course if he is your best friend and has saved yours a number of times; you do it without even thinking. The situation is entirely different if the man for whose life you are expected to risk your own or for whom you are supposed to make other sacrifices, is an anonymous contemporary on whom you have never set eyes. In this case it is not love for the fellow human being that activates self-denying behavior—if indeed it is activated—but the love for some culturally evolved traditional norm of social behavior. Love of something or other is, in very many cases, the motivation behind the power of the categorical imperative—an assertion which, I think, Kant would deny.

Our Cro-Magnon warrior had plenty of hostile neighbors against whom to discharge his aggressive drive, and he had just the right number of reliable friends to love. His moral responsibility was not overtaxed by an exercise of function which prevented him from striking, in sudden anger, at his companions with his sharpened hand-ax. The increase in number of individuals belonging to the same community is in itself sufficient to upset the balance between the personal bonds and the aggressive drive. It is definitely detrimental to the bond of friendship if a person has too many friends. It is proverbial that one can have only a few really close friends. To have a large number of "acquaintances," many of whom may be faithful allies with a legitimate claim to be regarded as real friends, overtaxes a man's capacity for personal love and dilutes the intensity of his emotional attachment. The close crowding of many individuals in a small space brings about a fatigue of all social reactions. Every inhabitant of a modern city is familiar with the surfeit of social relationships and responsibilities and knows the disturbing feeling of not being as pleased as he ought to be at the visit of a friend, even if he is genuinely fond of him and has not seen him for a long time. A tendency to bad temper is experienced when the telephone rings after dinner. That crowding increases the propensity to aggressive behavior has long been known and demonstrated experimentally by sociological research.

On the other hand, there is, in the modern community, no legitimate outlet for aggressive behavior. To keep the peace is the first of civic duties, and the hostile neighboring tribe, once the target at which to discharge phylogenetically programmed aggression, has now withdrawn to an ideal distance, hidden behind a curtain, if possible of iron. Among the many phylogenetically adapted norms of human social behavior, there is hardly one that does not need to be controlled and kept on a leash by responsible morality. This indeed is the deep truth contained in all sermons preaching asceticism. Most of the vices and mortal sins condemned today correspond to inclinations that were purely adaptive or at least harmless in primitive man. Paleolithic people hardly ever had enough to eat and if, for once, they had trapped a mammoth, it was biologically correct and moral for every member of the horde to gorge to his utmost capacity; gluttony was not a vice. When, for once, they were fully fed, primitive human beings rested from their strenuous life and were as absolutely lazy as possible, but there was nothing reprehensible in their sloth. Their life was so hard that there was no danger of healthy sensuality degenerating into debauch. A man sorely needed to keep his few possessions, weapons and tools, and a few nuts for tomorrow's meal; there was no danger of his hoarding instinct turning into avarice. Alcohol was not invented, and there are no indications that man had discovered the reinforcing properties of alkaloids, the only real vices known of present-day primitive tribes. In short, man's endowment with phylogenetically adapted patterns of behavior met the requirements well enough to make the task of responsible morality very easy indeed. Its only commandment at the time was: Thou shalt not strike thy neighbor with a hand-ax even if he angers thee.

Clearly, the task of compensation devolving on responsible morality increases at the same rate as the ecological and sociological conditions created by culture deviate from those to which human instinctive behavior is phylogenetically adapted. Not only does this deviation continue to increase, but it does so with an acceleration that is truly frightening.

The fate of humanity hangs on the question whether or not responsible morality will be able to cope with its rapidly growing burden. We shall not lighten this burden by overestimating the strength of morality, still less by attributing omnipotence to it. We have better chances of supporting moral responsibility in its ever-increasing task if we humbly realize and acknowledge that it is "only" a compensatory mechanism of very limited strength and that, as I have already explained, it derives what power it has from the same kind of motivational sources as those which it has been created to control. I have already said that the dynamics of instinctive drives, of phyletically and culturally ritualized behavior patterns, together with the controlling force of responsible morality, form a very complicated systemic whole which is not easy to analyze. However, the recognition of the mutual functional interdependence of its parts, even at the present incomplete stage of our knowledge, helps us to understand a number of phenomena which otherwise would remain completely unintelligible.

We all suffer to some extent from the necessity to control our natural inclinations by the exercise of moral responsibility. Some of us, lavishly endowed with social inclinations, suffer hardly at all; other less lucky ones need all the strength of their sense of moral responsibility to keep from getting into trouble with the strict requirements of modern society. According to a useful old psychiatric definition, a psychopath is a man who either suffers himself from the demands of society or else makes society suffer. Thus in one sense we are all psychopaths, for each of us suffers from the necessity of self-imposed control for the good of the community. The above-mentioned definition, however, was meant to apply particularly to those people who do not just suffer in secret, but overtly break down under the stress imposed upon them, becoming either neurotic or delinquent. Even according to this much narrower interpretation of our definition, the "normal" human being differs from the psychopath, the good man from the criminal, much less sharply than the healthy differs from the pathological. This difference is analogous to that between a man with a compensated valvular deficiency of the heart and one with a decompensated heart disease. In the first case, an increase of the work performed by the heart muscles is sufficient to compensate for the mechanical defect of the valve, so that the over-all pumping performance of the heart is adapted to the requirements of the body, at least for the time being. When the muscle finally breaks down under the prolonged strain, the heart becomes "decompensated." This analogy also goes to show that the compensatory function uses up energy.

This explanation of the essential function of responsible morality resolves a contradiction in Kant's doctrine of morality which was noticed earlier by Friedrich Schiller. He whom Herder called "the most inspired of all Kantians" opposed Kant's devaluation of all natural inclinations and satirized it in the wonderful Xenie: *"Gerne dien' ich dem Freund, doch leider tu' ich's aus Neigung, darum wurmt es mich oft, dass ich nicht tugendhaft bin"*—"I like serving my friend but alas, I do it from inclination, and thus it often vexes me that I am not virtuous."

However, not only do we serve our friend by inclination but we judge his acts of friendship according to whether it was warm, natural inclination that prompted him to perform them. If we were utterly logical Kantians, we would have to do the opposite and value most the man who instinctively dislikes us but who by responsible self-questioning is forced, much against his inclinations, to treat us kindly; however, in actual fact we can feel at most a tepid form of respect for such a benefactor, but we have a warm affection for the man who treats us as a friend because he "feels that way," without thinking that he is doing something worthy of gratitude.

When my unforgettable teacher, Ferdinand Hochstetter, at the age of seventy-one gave his valedictory address at Vienna University, the then Chancellor thanked him warmly for his long and inspired work. Hochstetter's answer put in a nutshell the whole paradox of value and nonvalue of natural inclination. This is what he said: "You are thanking me for something for which I deserve no gratitude. Thank my parents, my ancestors who transmitted to me these and no other inclinations. And if you ask me what I have done throughout my life in the fields of research and teaching then I must honestly say: I have always done the thing which, at the moment, I considered the greatest fun!"

What a strange contradiction! This great scientist who, as I know for a fact, had never read Kant, here shared the philosopher's standpoint in denying all value to natural inclination while, at the same time, the inestimable value of his work, accomplished "just for fun," reduces the Kantian theory of values and morality *ad absurdum* even more effectively than Friedrich Schiller's succinct stanza.

Yet it is easy to resolve this seeming contradiction, if we keep in mind that moral responsibility functions, as a compensatory mechanism, in a system of which natural inclination, by no means necessarily devoid of value, forms another indispensable part.

If we are assessing the behavior of a certain person—of ourselves, for example—we will naturally rate any particular action the higher the less it is motivated by natural inclination.

On the other hand, if we are assessing people as friends, we will naturally prefer the one whose friendship does not stem from rational considerations—however moral these may be—but from the warm feelings of natural inclination. It is no paradox but plain common sense that we use two different standards for judging the deeds of a man and the man himself.

The man who behaves socially from natural inclination normally makes few demands on the controlling mechanism of his own moral responsibility. Thus, in times of stress, he has huge reserves of moral strength to draw upon. But the man who, even in everyday life, has constantly to exert all his moral strength in order to curb his natural inclination into a semblance of normal social behavior, is very likely to break down completely in case of additional stress. Our parable of the compensated heart disorder applies quite exactly here, particularly regarding its energetical aspects.

The stress under which morally responsible behavior breaks down can be of varying kinds. It is not so much the sudden, one-time great temptation that makes human morality break down but the effect of any prolonged situation that exerts an increasing drain on the compensatory power of morality. Hunger, anxiety, the necessity to make difficult decisions, overwork, hopelessness and the like all have the effect of sapping moral energy and, in the long run, making it break down. Anyone who has had the opportunity to observe men under this kind of strain, for example in war or in prisoner-of-war camps, knows how unpredictably and suddenly moral decompensation sets in. Men in whose strength one trusted unconditionally suddenly break down, and others of whom one would never have expected it prove to be sources of inexhaustible energy, keeping up the morale of others by their example. Anyone who has experienced such things knows that the fervor of good intention and its power of endurance are two independent variables. Once you have realized this, you cease to feel superior to the man who breaks down a little sooner than you do yourself. Even the best and noblest reaches a point where his resistance is at an end: *"Eloi, Eloi, lama sabachthani?"*

As already mentioned, norms of social behavior developed by cultural ritualization play at least as important a part in the context of human society as instinctive motivation and the control exerted by responsible morality. Even at the earliest dawn of culture, when the invention of tools was just beginning to upset the equilibrium of phylogenetically evolved patterns of social behavior, man's newborn responsibility must have found a strong aid in cultural ritualization. Evidence of cultural rites reaches back almost as far as that of the use of tools and of fire. Of course we can expect prehistorical evidence of culturally ritualized behavior only when ritualization has reached comparatively high levels of differentiation, as in burial ceremonies or in the arts of painting and sculpture. These make their first appearance simultaneously with our own species, and the marvelous proficiency of the first known painters and sculptors suggests that even by their time, art had quite a long history behind it. Considering all this, it is quite possible that a cultural tradition of behavioral norms originated as early as the use of tools or even earlier. The beginnings of both have been found in the chimpanzee.

Through the processes described in Chapter Five, customs and taboos may acquire the power to motivate behavior in a way comparable to that of autonomous instincts. Not only highly developed rites or ceremonies but also simpler and less conspicuous norms of social behavior may attain, after a number of generations, the character of sacred customs which are loved and considered as values whose infringement is severely frowned upon by public opinion. As also has already been hinted in Chapter Five, sacred custom owes its motivating force to phylogenetically evolved behavior patterns of which two are of particular importance. One is response of militant enthusiasm by which any group defends its own social norms and rites against another group not possessing them; the other is the group's cruel taunting of any of its members who fail to conform with the accepted "good form" of behavior. Without the phylogenetically programmed love for traditional custom, human society would lack the supporting apparatus to which its owes its indispensable structure. Yet, like any phylogenetically programmed behavior mechanism, the one under discussion can miscarry. School classes or companies of soldiers, both of which can be regarded as models of primitive group structure, can be very cruel indeed in their ganging up against an outsider. The purely instinctive response to a physically abnormal individual, for instance the jeering at a fat boy, is, as far as overt behavior is concerned, absolutely identical with discrimination against a person who differs from the group in culturally developed social norms—for instance, a child who speaks a different dialect.

The ganging up on an individual diverging from the social norms characteristic of a group and the group's enthusiastic readiness to defend these social norms and rites are both good illustrations of the way in which culturally determined conditioned-stimulus situations release activities which are fundamentally instinctive. They are also excellent examples of typical compound behavior patterns whose primary survival value is as obvious as the danger of their misfiring under the conditions of the modern social order. I shall have to come back to the different ways in which the function of militant enthusiasm can miscarry and to possible means of preventing this eventuality.

Before enlarging on this subject, however, a few words must be said about the functions of social norms and rites in general. First of all I must recall to the reader's memory the somewhat surprising fact mentioned in Chapter Five: We have no immediate knowledge of the function and/or survival value of the majority of our own established customs, notwithstanding our emotional conviction that they do indeed constitute high values. This paradoxical state of affairs is explained by the simple fact that customs are not man-made in the same sense as human inventions are, from the pebble tool up to the jet plane.

There may be exceptional cases in which causal insight gained by a great lawgiver determines a social norm. Moses is said to have recognized the pig as a host of the Trichina, but if he did, he preferred to rely on the devout religious observance of his people rather than on their intellect when he asserted that Jehovah himself had declared the porker an unclean animal. In general, however, it is quite certain that it hardly ever was insight into a valuable function that gave rise to traditional norms and rites, but the age-old process of natural selection. Historians will have to face the fact that natural selection determined the evolution of cultures in the same manner as it did that of species.

In both cases, the great constructor has produced results which may not be the best of all conceivable solutions but which at least prove their viability by their very existence. To the biologist who knows the ways in which selection works and who is also aware of its limitations it is in no way surprising to find, in its constructions, some details which are unnecessary or even detrimental to survival. The human mind, endowed with the power of deduction, can quite often find solutions to problems which natural selection fails to resolve. Selection may produce incomplete adaptation even when it uses the material furnished by mutation and when it has at its disposal huge time periods. It is much more likely to do so when it has to determine, in an incomparably shorter time, which of the randomly arising customs of a culture make it best fitted to survival. Small wonder indeed if, among the social norms and rites of any culture, we find a considerable number which are unnecessary or even clearly inexpedient and which selection nevertheless has failed to eliminate. Many superstitions, comparable to my little greylag's detour toward the window, can become institutionalized and be carried on for generations. Also, intra-specific selection often plays as dangerous a role in the development of cultural ritualization as in phylogenesis. The process of so-called status-seeking, for instance, produces the bizarre excrescences in social norms and rites which are so typical of intra-specific selection.

However, even if some social norms or rites are quite obviously maladaptive, this does not imply that they may be eliminated without further consideration. The social organization of any culture is a complicated system of universal interaction between a great many divergent traditional norms of behavior, and it can never be predicted without a very thorough analysis what repercussions the cutting out of even one single part may have for the functioning of the whole. For instance, it is easily intelligible to anybody that the custom of head-hunting, widely spread among tropical tribes, has a somewhat unpleasant side to it, and that the peoples still adhering to it would be better off, in many ways, without it. The studies of the ethnologist and psychoanalyst Derek Freeman, however, have shown that head-hunting is so intricately interwoven with the whole social system of some Bornean tribes that its abolition tends to disintegrate their whole culture, even seriously jeopardizing the survival of the people.

The balanced interaction between all the single norms of social behavior characteristic of a culture accounts for the fact that it usually proves highly dangerous to mix cultures. To kill a culture, it is often sufficient to bring it into contact with another, particularly if the latter is higher, or is at least regarded as higher, as the culture of a conquering nation usually is. The people of the subdued side then tend to look down upon everything they previously held sacred and to ape the customs which they regard as superior. As the system of social norms and rites characteristic of a culture is always adapted, in many particular ways, to the special conditions of its environment, this unquestioning acceptance of foreign customs almost invariably leads to maladaptation. Colonial history offers abundant examples of its causing the destruction not only of cultures but also of peoples and races. Even in the less tragic case of rather closely related and roughly equivalent cultures mixing, there usually are some undesirable results, because each finds it easier to imitate the most superficial, least valuable customs of the other. The first items of American culture imitated by German youth immediately after the last war were gum chewing, Coca-Cola drinking, the crew cut, and the reading of color comic strips. More valuable social norms characteristic of American culture were obviously less easy to imitate.

Quite apart from the danger to one culture arising from contact with another, all systems of social norms and rites are vulnerable in the same way as systems of phylogenetically evolved patterns of social behavior. Not being man-made, but produced by selection, their function is, without special scientific investigation, unknown to man himself, and therefore their balance is as easily upset by the effects of conceptual thought as that of any system of instinctive behavior. Like the latter, they can be made to miscarry by any environmental change not "foreseen" in their "programming," but while instincts persist for better or worse, traditional systems of social behavior can disappear altogether within one generation, because, like the continuous state that constitutes the life of an organism, that which constitutes a culture cannot bear any interruption of its continuity.

Several coinciding factors are, at present, threatening to interrupt the continuity of our Western culture. There is, in our culture, an alarming break of traditional continuity between the generation born at about 1900 and the next. This fact is incontestable; its causes are still doubtful. Diminishing cohesion of the family group and decreasing personal contact between teacher and pupil are probably important factors. Very few of the present younger generation have ever had the opportunity of seeing their fathers at work; few pupils learn from their teachers by collaborating with them. This used to be the rule with peasants, artisans, and even scientists, provided they taught at relatively small universities. The industrialization that prevails in all sectors of human life produces a distance between the generations which is not compensated for by the greatest familiarity, by the most democratic tolerance and permissiveness of which we are so proud. Young people seem to be unable to accept the values held in honor by the older generation, unless they are in close contact with at least one of its representatives who commands their unrestricted respect and love.

Another probably important factor contributing to the same effect is the real obsolescence of many social norms and rites still valued by some of the older generation. The extreme speed of ecological and sociological change wrought by the development of technology causes many customs to become maladaptive within one generation. The romantic veneration of national values, so movingly expressed in the works of Rudyard Kipling or C. S. Forester, is obviously an anachronism that can do nothing but damage today.

Such criticism is indubitably overstressed by the prevalence of scientific thought and the unrelenting demand for causal understanding, both of which are the most characteristic, if not the only, virtues of our century. However, scientific enlightenment tends to engender doubt in the value of traditional beliefs long before it furnishes the causal insight necessary to decide whether some accepted custom is an obsolete superstition or a still indispensable part of a system of social norms. Again it is the unripe fruit of the tree of knowledge that proves to be dangerous; indeed, I suspect that the whole legend of the tree of knowledge is meant to defend sacred traditions against the premature inroads of incomplete rationalization.

As it is, we do not know enough about the function of any system of culturally ritualized norms of behavior to give a rational answer to the perfectly rational question of what some particular custom is good for, in other words wherein lies its survival value. When an innovator rebels against established norms of social behavior and asks why he should conform with them, we are usually at a loss for an answer. It is only in rare cases, as in my example of Moses' law against eating pigs, that we can give the would-be reformer such a succinct answer as: "You will get trichinosis if you don't obey." In most cases the defender of accepted tradition has to resort to seemingly lame replies, saying that certain things are "simply not done," are not cricket, are un-American or sinful, if he does not prefer to appeal to the authority of some venerable father-figure who also regarded the social norm under discussion as inviolable.

To anyone for whom the latter is still endowed with the emotional value of a sacred rite, such an answer appears as self-evident and satisfactory; to anybody who has lost this feeling of reverence it sounds hollow and sanctimonious. Understandably, if not quite forgivably, such a person tends to think that the social norm in question is just superstition, if he does not go so far as to consider its defender as insincere. This, incidentally, is very frequently the main point of dissension between people of different generations.

In order correctly to appreciate how indispensable cultural rites and social norms really are, one must keep in mind that, as Arnold Gehlen has put it, man is by nature a being of culture. In other words, man's whole system of innate activities and reactions is phylogenetically so constructed, so "calculated" by evolution, as to *need* to be complemented by cultural tradition. For instance, all the tremendous neurosensory apparatus of human speech is phylogenetically evolved, but so constructed that its function presupposes the existence of a culturally developed language which the infant has to learn. The greater part of all phylogenetically evolved patterns of human social behavior is interrelated with cultural tradition in an analogous way. The urge to become a member of a group, for instance, is certainly something that has been programmed in the prehuman phylogeny of man, but the distinctive properties of any group which make it coherent and exclusive are norms of behavior ritualized in cultural development. As has been explained in Chapter Five, without traditional rites and customs representing a common property valued and defended by all members of the group, human beings would be quite unable to form social units exceeding in size that of the primal family group which can be held together by the instinctive bond of personal friendship discussed in Chapter Eleven.

The equipment of man with phylogenetically programmed norms of behavior is just as dependent on cultural tradition and rational responsibility as, conversely, the function of both the latter is dependent on instinctual motivation (pages 77 f). Were it possible to rear a human being of normal genetic constitution under circumstances depriving it of all cultural tradition—which is impossible not only for ethical but also for biological reasons—the subject of the cruel experiment would be very far from representing a reconstruction of a prehuman ancestor, as yet devoid of culture. It would be a poor cripple, deficient in higher functions in a way comparable to that in which idiots who have suffered encephalitis during infantile or fetal life lack the higher functions of the cerebral cortex. No man, not even the greatest genius, could invent, all by himself, a system of social norms and rites forming a substitute for cultural tradition.

In our time, one has plenty of unwelcome opportunity to observe the consequences which even a partial deficiency of cultural tradition has on social behavior. The human beings thus affected range from young people advocating necessary if dangerous abrogations of customs that have become obsolete, through angry young men and rebellious gangs of juveniles, to the appearance of a certain well-defined type of juvenile delinquent which is the same all over the world. Blind to all values, these unfortunates are the victims of infinite boredom.

The means by which an expedient compromise between the rigidity of social norms and the necessity of adaptive change can be effected is prescribed by biological laws of the widest range of application. No organic system can attain to any higher degree of differentiation without firm and cohesive structures supporting it and holding it together. Such a structure and its support can, in principle, only be gained by the sacrifice of certain degrees of freedom that existed before. A worm can bend all over, an arthropod only where its cuticular skeleton is provided with joints for that purpose.

Changes in outer or inner environment may demand degrees of freedom not permitted by the existing structure and therefore may necessitate its partial and/or temporary disintegration, in the same way that growth necessitates the periodic shedding of the shell in crustacea and other arthropods. This act of demolishing carefully erected structures, though indispensable if better adapted ones are to arise, is always followed by a period of dangerous vulnerability, as is impressively illustrated by the defenseless situation of the newly molted soft-shelled crab.

All this applies unrestrictedly to the "solidified," that is to say institutionalized, system of social norms and rites which function very much like a supporting skeleton in human cultures. In the growth of human cultures, as in that of arthropods, there is a built-in mechanism providing for graduated change. During and shortly after puberty human beings have an indubitable tendency to loosen their allegiance to all traditional rites and social norms of their culture, allowing conceptual thought to cast doubt on their value and to look around for new and perhaps more worthy ideals. There probably is, at that time of life, a definite sensitive period for a new object-fixation, much as in the case of the object-fixation found in animals and called imprinting. If at that critical time of life old ideals prove fallacious under critical scrutiny and new ones fail to appear, the result is complete aimlessness, the utter boredom which characterizes the young delinquent. If, on the other hand, the clever demagogue, well versed in the dangerous art of producing supranormal stimulus situations, gets hold of young people at the susceptible age, he finds it easy to guide their object-fixation in a direction subservient to his political aims. At the postpuberal age some human beings seem to be driven by an overpowering urge to espouse a cause and failing to find a worthy one may become fixated on astonishingly inferior substitutes. The instinctive need to be the member of a closely knit group fighting for common ideals may grow so strong that it becomes inessential what these ideals are and whether they possess any intrinsic value. This, I believe, explains the formation of juvenile gangs whose social structure is very probably a rather close reconstruction of that prevailing in primitive human society.

Apparently this process of object-fixation can take its full effect only once in an individual's life. Once the valuation of certain social norms or the allegiance to a certain cause is fully established, it cannot be erased again, at least not to the extent of making room for a new, equally strong one. Also it would seem that once the sensitive period has elapsed, a man's ability to embrace ideals at all is considerably reduced. All this helps to explain the hackneyed truth that human beings have to live through a rather dangerous period at, and shortly after, puberty. The tragic paradox is that the danger is greatest for those who are by nature best fitted to serve the noble cause of humanity.

The process of object-fixation has consequences of an importance that can hardly be overestimated. It determines neither more nor less than that which a man will live for, struggle for, and, under certain circumstances, blindly go to war for. It determines the conditioned stimulus situation releasing a powerful phylogenetically evolved behavior which I propose to call that of militant enthusiasm.

Militant enthusiasm is particularly suited for the paradigmatic illustration of the manner in which a phylogenetically evolved pattern of behavior interacts with culturally ritualized social norms and rites, and in which, though absolutely indispensable to the function of the compound system, it is prone to miscarry most tragically if not strictly controlled by rational responsibility based on causal insight. The Greek word *enthousiasmos* implies that a person is possessed by a god; the German *Begeisterung* means that he is controlled by a spirit, a *Geist,* more or less holy.

In reality, militant enthusiasm is a specialized form of communal aggression, clearly distinct from and yet functionally related to the more primitive forms of petty individual aggression. Every man of normally strong emotions knows, from his own experience, the subjective phenomena that go hand in hand with the response of militant enthusiasm. A shiver runs down the back and, as more exact observation shows, along the outside of both arms. One soars elated, above all the ties of everyday life, one is ready to abandon all for the call of what, in the moment of this specific emotion, seems to be a sacred duty. All obstacles in its path become unimportant; the instinctive inhibitions against hurting or killing one's fellows lose, unfortunately, much of their power. Rational considerations, criticism, and all reasonable arguments against the behavior dictated by militant enthusiasm are silenced by an amazing reversal of all values, making them appear not only untenable but base and dishonorable. Men may enjoy the feeling of absolute righteousness even while they commit atrocities. Conceptual thought and moral responsibility are at their lowest ebb. As a Ukrainian proverb says: "When the banner is unfurled, all reason is in the trumpet."

The subjective experiences just described are correlated with the following, objectively demonstrable phenomena. The tone of the entire striated musculature is raised, the carriage is stiffened, the arms are raised from the sides and slightly rotated inward so that the elbows point outward. The head is proudly raised, the chin stuck out, and the facial muscles mime the "hero face," familiar from the films. On the back and along the outer surface of the arms the hair stands on end. This is the objectively observed aspect of the shiver!

Anybody who has ever seen the corresponding behavior of the male chimpanzee defending his band or family with self-sacrificing courage will doubt the purely spiritual character of human enthusiasm. The chimp, too, sticks out his chin, stiffens his body, and raises his elbows; his hair stands on end, producing a terrifying magnification of his body contours as seen from the front. The inward rotation of his arms obviously has the purpose of turning the longest-haired side outward to enhance the effect. The whole combination of body attitude and hair-raising constitutes a bluff. This is also seen when a cat humps its back, and is calculated to make the animal appear bigger and more dangerous than it really is. Our shiver, which in German poetry is called a *"heiliger Schauer,"* a "holy" shiver, turns out to be the vestige of a prehuman vegetative response of making a fur bristle which we no longer have.

To the humble seeker of biological truth there cannot be the slightest doubt that human militant enthusiasm evolved out of a communal defense response of our prehuman ancestors. The unthinking single-mindedness of the response must have been of high survival value even in a tribe of fully evolved human beings. It was necessary for the individual male to forget all his other allegiances in order to be able to dedicate himself, body and soul, to the cause of the communal battle. *"Was schert mich Weib, was schert mich Kind"*—"What do I care for wife or child," says the Napoleonic soldier in a famous poem by Heinrich Heine, and it is highly characteristic of the reaction that this poet, otherwise a caustic critic of emotional romanticism, was so unreservedly enraptured by his enthusiasm for the "great" conqueror as to find this supremely apt expression.

The object which militant enthusiasm tends to defend has changed with cultural development. Originally it was certainly the community of concrete, individually known members of a group, held together by the bond of personal love and friendship. With the growth of the social unit, the social norms and rites held in common by all its members became the main factor holding it together as an entity, and therewith they became automatically the symbol of the unit. By a process of true Pavlovian conditioning plus a certain amount of irreversible imprinting these rather abstract values have in every human culture been substituted for the primal, concrete object of the communal defense reaction.

This traditionally conditioned substitution of object has important consequences for the function of militant enthusiasm. On the one hand, the abstract nature of its object can give it a definitely inhuman aspect and make it positively dangerous—what do I care for wife or child; on the other hand it makes it possible to recruit militant enthusiasm in the service of really ethical values. Without the concentrated dedication of militant enthusiasm neither art, nor science, nor indeed any of the great endeavors of humanity would ever have come into being. Whether enthusiasm is made to serve these endeavors, or whether man's most powerfully motivating instinct makes him go to war in some abjectly silly cause, depends almost entirely on the conditioning and/or imprinting he has undergone during certain susceptible periods of his life. There is reasonable hope that our moral responsibility may gain control over the primeval drive, but our only hope of its ever doing so rests on the humble recognition of the fact that militant enthusiasm is an instinctive response with a phylogenetically determined releasing mechanism and that the only point at which intelligent and responsible supervision can get control is in the conditioning of the response to an object which proves to be a genuine value under the scrutiny of the categorical question.

Like the triumph ceremony of the greylag goose, militant enthusiasm in man is a true autonomous instinct: it has its own appetitive behavior, its own releasing mechanisms, and, like the sexual urge or any other strong instinct, it engenders a specific feeling of intense satisfaction. The strength of its seductive lure explains why intelligent men may behave as irrationally and immorally in their political as in their sexual lives. Like the triumph ceremony, it has an essential influence on the social structure of the species. Humanity is not enthusiastically combative because it is split into political parties, but it is divided into opposing camps because this is the adequate stimulus situation to arouse militant enthusiasm in a satisfying manner. "If ever a doctrine of universal salvation should gain ascendancy over the whole earth to the exclusion of all others," writes Erich von Holst, "it would at once divide into two strongly opposing factions (one's own true one and the other heretical one) and hostility and war would thrive as before, mankind being—unfortunately—what it is!"

The first prerequisite for rational control of an instinctive behavior pattern is the knowledge of the stimulus situation which releases it. Militant enthusiasm can be elicited with the predictability of a reflex when the following environmental situations arise. First of all, a social unit with which the subject identifies himself must appear to be threatened by some danger from outside. That which is threatened may be a concrete group of people, the family or a little community of close friends, or else it may be a larger social unit held together and symbolized by its own specific social norms and rites. As the latter assume the character of autonomous values, in the way described in Chapter Five, they can, quite by themselves, represent the object in whose defense militant enthusiasm can be elicited. From all this it follows that this response can be brought into play in the service of extremely different objects, ranging from the sports club to the nation, or from the most obsolete mannerisms or ceremonials to the ideal of scientific truth or of the incorruptibility of justice.

A second key stimulus which contributes enormously to the releasing of intense militant enthusiasm is the presence of a hated enemy from whom the threat to the above "values" emanates. This enemy, too, can be of a concrete or of an abstract nature. It can be "the" Jews, Huns, Boches, tyrants, etc., or abstract concepts like world capitalism, Bolshevism, fascism, and any other kind of ism; it can be heresy, dogmatism, scientific fallacy, or what not. Just as in the case of the object to be defended, the enemy against whom to defend it is extremely variable, and demagogues are well versed in the dangerous art of producing supranormal dummies to release a very dangerous form of militant enthusiasm.

A third factor contributing to the environmental situation eliciting the response is an inspiring leader figure. Even the most emphatically antifascistic ideologies apparently cannot do without it, as the giant pictures of leaders displayed by all kinds of political parties prove clearly enough. Again the un-selectivity of the phylogenetically programmed response allows for a wide variation in the conditioning to a leader figure. Napoleon, about whom so critical a man as Heinrich Heine became so enthusiastic, does not inspire me in the least; Charles Darwin does.

A fourth, and perhaps the most important, prerequisite for the full eliciting of militant enthusiasm is the presence of many other individuals, all agitated by the same emotion. Their absolute number has a certain influence on the quality of the response. Smaller numbers at issue with a large majority tend to obstinate defense with the emotional value of "making a last stand," while very large numbers inspired by the same enthusiasm feel the urge to conquer the whole world in the name of their sacred cause. Here the laws of mass enthusiasm are strictly analogous to those of flock formation described in Chapter Eight; here, too, the excitation grows in proportion, perhaps even in geometrical progression, with the increasing number of individuals. This is exactly what makes militant mass enthusiasm so dangerous.

I have tried to describe, with as little emotional bias as possible, the human response of enthusiasm, its phylogenetic origin, its instinctive as well as its traditionally handed-down components and prerequisites. I hope I have made the reader realize, without actually saying so, what a jumble our philosophy of values is. What is a culture? A system of historically developed social norms and rites which are passed on from generation to generation because emotionally they are felt to be values. What is a value? Obviously, normal and healthy people are able to appreciate something as a high value for which to live and, if necessary, to die, for no other reason than that it was evolved in cultural ritualization and handed down to them by a revered elder. Is, then, a value only defined as the object on which our instinctive urge to preserve and defend traditional social norms has become fixated? Primarily and in the early stages of cultural development this indubitably was the case. The obvious advantages of loyal adherence to tradition must have exerted a considerable selection pressure. However, the greatest loyalty and obedience to culturally ritualized norms of behavior must not be mistaken for responsible morality. Even at their best, they are only functionally analogous to behavior controlled by rational responsibility. In this respect, they are no whit different from the instinctive patterns of social behavior discussed in Chapter Seven. Also they are just as prone to miscarry under circumstances for which they have not been "programmed" by the great constructor, natural selection.

In other words, the need to control, by wise rational responsibility, all our emotional allegiances to cultural values is as great as, if not greater than, the necessity to keep in check our other instincts. None of them can ever have such devastating effects as unbridled militant enthusiasm when it infects great masses and overrides all other considerations by its single-mindedness and its specious nobility. It is not enthusiasm in itself that is in any way noble, but humanity's great goals which it can be called upon to defend. That indeed is the Janus head of man: The only being capable of dedicating himself to the very highest moral and ethical values requires for this purpose a phylogenetically adapted mechanism of behavior whose animal properties bring with them the danger that he will kill his brother, convinced that he is doing so in the interests of these very same high values. *Ecce homo!*

Chapter Fourteen

Avowal of Optimism

Ich bilde mir nicht ein, ich könnte was lehren
Die Menschen zu bessern und zu bekehren.—GOETHE *

I do not mind admitting that, unlike Faust, I think I have something to teach mankind that may help it to change for the better. This conviction is not as presumptuous as it might seem; it is certainly less so than the opposite stand, which is usually based not so much on a man's distrust of his capacity to teach as on the haughty assumption that humanity is not ready to understand the profound truths of his new doctrine. This is true only in those very rare instances when an intellectual giant is centuries ahead of his time. He is misunderstood and runs the risk either of beoming a martyr or of being brushed aside as a crank. If his contemporaries pay attention to a teacher or even read his books, it can safely be assumed that he is not an intellectual giant. At best he can flatter himself that he has something to say that is "due" to be said at that moment. His teachings will be most efficacious if his ideas are only a short head in front of his hearers. A new truth has really convinced when the hearer exclaims, "How silly of me

* "I do not set myself up as having anything to teach mankind to help it toward conversion or self-improvement."

not to have thought of that," as Thomas Huxley is reported to have said on reading Charles Darwin's *Origin of Species.*

I am really being far from presumptuous when I profess my conviction that in the very near future not only scientists, but the majority of tolerably intelligent people, will consider as an obvious and banal truth all that has been said in this book about instincts in general and intra-specific aggression in particular, about phylogenetic and cultural ritualization, and about the factors that build up the ever-increasing danger of human society's becoming completely disintegrated by the misfunctioning of social behavior patterns.

There is less hazard of my meeting with disbelief than of incurring the reproach of banality when I now proceed to summarize the most important inferences from what has been said in this book by formulating simple precepts for preventive measures against that danger. I am aware that these measures must appear feeble and ineffective after all I have said in the last chapter about the present situation of mankind. This, however, does not argue against the correctness of my inferences. In medicine, too, all therapeutic measures appear slight and ineffectual when compared with the amount of physiological and pathological knowledge and insight which had to be gained before any reasonable therapy at all could be planned. Science seldom effects dramatic changes in the course of history, except, of course, in the sense of destruction, for it is all too easy to misuse the power afforded by causal insight. To use the knowledge gained by scientific research in a creative and beneficial fashion demands no less perspicacity and meticulous application to detail than were necessary to gain it.

The first, the most obvious, and the most important precept is the old Τνῶθι σεαυτὸν, "Know thyself": we must deepen our insight into the causal concatenations governing our own behavior. The lines along which an applied science of human behavior will probably develop are just beginning to appear.

One line is the objective, ethological investigation of all the possibilities of discharging aggression in its primal form on substitute objects, and we already know that there are better ones than to kick empty carbide tins. The second is the psychoanalytic study of so-called sublimation. We may anticipate that a deeper knowledge of this specifically human form of catharsis will do much toward the relief of undischarged aggressive drives. The third way of avoiding aggression, though an obvious one, is still worth mentioning: it is the promotion of personal acquaintance and, if possible, friendship between individual members of different ideologies or nations. The fourth and perhaps the most important measure to be taken immediately is the intelligent and responsible channeling of militant enthusiasm, in other words helping a younger generation which, on the one hand, is highly critical and even suspicious and, on the other, emotionally starved, to find genuine causes that are worth serving in the modern world. I shall now discuss all these precepts one by one.

Even at its present modest stage, our knowledge of the nature of aggression is sufficient to tell us what measures against its damaging effects have no hope of success whatever, and this in itself is of value. To anybody who is unaware of the essential spontaneity of instinctive drives and who is wont to think of behavior exclusively in terms of conditioned and unconditioned responses, it must seem a hopeful undertaking to diminish or even eliminate aggression by shielding mankind from all stimulus situations eliciting aggressive behavior. The results of this experiment have already been discussed in Chapter Four. Another unpromising attempt is to control aggression by putting a moral veto on it. The practical application of both these methods would be about as judicious as trying to counteract the increasing pressure in a continuously heated boiler by screwing down the safety valve more tightly.

A further, theoretically possible but in my opinion highly

inadvisable measure would be to attempt to breed out the aggressive drive by eugenic planning. We know from the preceding chapters that there is intra-specific aggression in the human reaction of enthusiasm and this, though dangerous, is nevertheless indispensable for the achievement of the highest human goals. We know from the chapter "The Bond" that aggression in very many animals and probably also in man is an essential component of personal friendship. Finally, in the chapter on the great parliament of instincts, we have learned how complex is the interaction of different drives. It would have quite unpredictable consequences if one of them—and one of the strongest—were to disappear entirely. We do not know how many important behavior patterns of man include aggression as a motivating factor, but I believe it occurs in a great many. What is certain is that, with the elimination of aggression, the *"aggredi"* in the original and widest sense, the tackling of a task or problem, the self-respect without which everything that a man does from morning till evening, from the morning shave to the sublimest artistic or scientific creations, would lose all impetus; everything associated with ambition, ranking order, and countless other equally indispensable behavior patterns would probably also disappear from human life. In the same way, a very important and specifically human faculty would probably disappear too: laughter.

The most promising means we can apply in our attempt to cope with the miscarrying of aggression—and that of other patterns of social behavior—are those which have proved their efficiency in the course of phylogenetic and cultural evolution.

A simple and effective way of discharging aggression in an innocuous manner is to redirect it at a substitute object. As explained in Chapter Eleven, this method has been employed extensively by the great constructors of evolution to prevent combat between members of a group. It is sound reason for optimism that aggression, more easily than most other instincts, can find complete satisfaction with substitute objects. Even without insight into the consequences of dammed-up drives, the choice of object is directed by reasonable considerations. I have found that even highly irascible people who, in a rage, seem to lose all control of their actions, still refrain from smashing really valuable objects, preferring cheaper crockery. Yet it would be a complete error to suspect that they could, if they only tried hard enough, keep from smashing things altogether! Insight into the physiology of dammed-up drive and its redirected discharge is, of course, a great help in governing aggression. It was certainly thanks to this insight that, in the incident related in Chapter Four, I did not hit my friend but jumped on an empty carbide tin; conversely, the reason that my old aunt, described in the same chapter, was so completely convinced of the depravity of her unfortunate housemaid was simply that she knew nothing of these phenomena. My dear old aunt was emphatically not my inferior in respect to moral self-control. Thus the differences in our behavior furnish a striking illustration of the fact that insight into the causality of our actions may endow our moral responsibility with the power to control them, even where the categorical imperative is doomed to fail miserably without that knowledge.

Redirection as a means of controlling the functions of aggression and other undischarged drives has been known to humanity for a long time. The ancient Greeks were familiar with the conception of catharsis, of purifying discharge, and psychoanalysis has shown very convincingly that many patterns of altogether laudable behavior derive their impulses from the "sublimation" of aggressive or sexual drives. Sublimation, however, must not be confounded with simple redirection of an instinctive activity toward a substitute object. There is a substantial difference between the man who bangs the table instead of hitting his antagonist, and the man who discharges the aggression aroused by an irritating family life by writing an enthusiastic pamphlet serving an altogether unconnected cause.

One of the many instances in which phylogenetic and cultural ritualization have hit on very similar solutions of the same problem concerns the method by which both have achieved the difficult task of avoiding killing without destroying the important functions performed by fighting in the interest of the species. All the culturally evolved norms of "fair fighting," from primitive chivalry to the Geneva Conventions, are functionally analogous to phylogenetically ritualized combat in animals.

It was probably in highly ritualized but still serious hostile fighting that sport had its origin. It can be defined as a specifically human form of nonhostile combat, governed by the strictest of culturally developed rules. Sport is not directly comparable to the fighting play of the higher vertebrates. The latter is never competitive, being essentially free from any appetitive or purposive tension. The enjoyable play of two dogs, however different in size and strength, is made possible only by the strict exclusion of all competitive elements. In sport, on the other hand, even in those kinds in which the enjoyment of skilled movements for their own sake predominates, as in skiing or skating, there always is a certain pride in doing it well, and there is no sport in which contests are not held. In this respect, human sport is more akin to serious fighting than animal play is; also, sport indubitably contains aggressive motivation, demonstrably absent in most animal play.

While some early forms of sport, like the jousting of medieval knights, may have had an appreciable influence on sexual selection, the main function of sport today lies in the cathartic discharge of aggressive urge; besides that, of course, it is of the greatest importance to keeping people healthy.

The value of sport, however, is much greater than that of a

simple outlet of aggression in its coarser and more individualistic behavior patterns, such as pummeling a punch-ball. It educates man to a conscious and responsible control of his own fighting behavior. Few lapses of self-control are punished as immediately and severely as loss of temper during a boxing bout. More valuable still is the educational value of the restrictions imposed by the demands for fairness and chivalry which must be respected even in the face of the strongest aggression-eliciting stimuli.

The most important function of sport lies in furnishing a healthy safety valve for that most indispensable and, at the same time, most dangerous form of aggression that I have described as collective militant enthusiasm in the preceding chapter. The Olympic Games are virtually the only occasion when the anthem of one nation can be played without arousing any hostility against another. This is so because the sportsman's dedication to the international social norms of his sport, to the ideals of chivalry and fair play, is equal to any national enthusiasm. The team spirit inherent in all international sport gives scope to a number of truly valuable patterns of social behavior which are essentially motivated by aggression and which, in all probability, evolved under the selection pressure of tribal warfare at the very dawn of culture. The noble warrior's typical virtues, such as his readiness to sacrifice himself in the service of a common cause, disciplined submission to the rank order of the group, mutual aid in the face of deadly danger, and, above all, a superlatively strong bond of friendship between men, were obviously indispensable if a small tribe of the type we have to assume for early man was to survive in competition with others. All these virtues are still desirable in modern man and still command our instinctive respect. It is undeniable that there is no situation in which all these virtues shine so brilliantly as they do in war, a fact which is dangerously liable to convince quite excellent but naïve people that war, after all, cannot be the absolutely abhorrent thing it really is.

Fortunately, there are other ways in which the above-mentioned, admittedly valuable virtues can be cultivated. The harder and more dangerous forms of sport, particularly those demanding the working together of larger groups, such as mountain climbing, diving, off-shore and ocean sailing, but also other dangerous undertakings, like polar expeditions and, above all, the exploration of space, all give scope for militant enthusiasm, allowing nations to fight each other in hard and dangerous competition without engendering national or political hatred. On the contrary, I am convinced that of all the people on the two sides of the great curtain, the space pilots are the least likely to hate each other. Like the late Erich von Holst, I believe that the tremendous and otherwise not quite explicable public interest in space flight arises from the subconscious realization that it helps to preserve peace. May it continue to do so!

Sporting contests between nations are beneficial not only because they provide an outlet for the collective militant enthusiasm of nations, but also because they have two other effects that counter the danger of war: they promote personal acquaintance between people of different nations or parties and they unite, in enthusiasm for a common cause, people who otherwise would have little in common. We must now discuss how these two measures against aggression work, and by what means they can be exploited to serve our purpose.

I have already said that we can learn much from demagogues who pursue the opposite purpose, namely to make peoples fight. They know very well that personal acquaintance, indeed every kind of brotherly feeling for the people to be attacked, constitutes a strong obstacle to aggression. Every militant ideology in history has propagated the belief that the members of the other party are not quite human, and every strategist is intent on preventing any "fraternization" between the soldiers in confronting trenches. Anonymity of the person to be attacked greatly facilitates the releasing of aggressive behavior. It is an observation familiar to anybody who has traveled in trains that well-bred people can behave atrociously toward strangers in the territorial defense of their compartment. When they discover that the intruder is an acquaintance, however casual, there is an amazing and ridiculous switch in their behavior from extreme rudeness to exaggerated and shamefaced politeness. Similarly, a naïve person can feel quite genuine hatred for an anonymous group, against "the" Germans, "the" Catholic foreigners, etc., and may rail against them in public, but he will never dream of being so much as impolite when he comes face to face with an individual member. On closer acquaintance with one or more members of the abhorred group such a person will rarely revise his judgment on it as a whole, but will explain his sympathy for individuals by the assumption that they are exceptions to the rule.

If mere acquaintance has this remarkable and altogether desirable effect, it is not surprising that real friendship between individuals of different nationality or ideology are even more beneficial. No one is able to hate, wholeheartedly, a nation among whose numbers he has several friends. Being friends with a few "samples" of another people is enough to awaken a healthy mistrust of all those generalizations which brand "the" Russians, English, Germans, etc., with typical and usually hateful national characteristics. To the best of my belief, my friend Walter Robert Corti was the first to put into practice the method of subduing international hatred by promoting international friendships. In his famous children's village in Trogen, Switzerland, children and young people of all kinds of nations are living together in a friendly community. May this attempt find imitators on a grand scale!

What is needed is the arousal of enthusiasm for causes

which are commonly recognized as values of the highest order by all human beings, irrespective of their national, cultural, or political allegiances. I have already called attention to the danger of defining a value by begging the question. A value is emphatically not just the object to which the instinctive response of militant enthusiasm becomes fixated by imprinting and early conditioning, even if, conversely, militant enthusiasm can become fixated on practically any institutionalized social norm or rite and make it appear as a value. Emotional loyalty to an institutionalized norm does not make it a value, otherwise war, even modern technical war, would be one. J. Marmor has quite recently called attention to the fact that even today "the history books of every nation justify its wars as brave, righteous, and honorable. This glorification is charged with overtones of patriotism and love of country. Virtues such as heroism and courage are regarded as being 'manly' and are traditionally associated with waging war. Conversely, the avoidance of war or the pursuit of peace are generally regarded as 'effeminate,' passive, cowardly, weak, dishonorable or subversive. The brutal realities, even of traditional war, are glamorized and obscured by countless tales of heroism and glory, and the warnings of an occasional General Sherman that 'war is hell [and] its glory all moonshine' are disregarded." I could not agree more with Dr. Marmor when he discusses the psychological obstacles to the elimination of war as a social institution and counts among them the insidious effect of military toys and war games which all prepare the soil for a psychological acceptance of war and violence. I agree with Dr. Marmor's assertion that modern war has become an institution, and I share his optimism in believing that, being an institution, war can be abolished.

However, I think we must face the fact that militant enthusiasm has evolved from the hackle-raising and chin-protruding communal defense instinct of our prehuman ancestors and that the key stimulus situations which release it still bear all the earmarks of this origin. Among them, the existence of an enemy, against whom to defend cultural values, is still one of the most effective. Militant enthusiasm, in one particular respect, is dangerously akin to the triumph ceremony of geese and to analogous instinctive behavior patterns of other animals. The social bond embracing a group is closely connected with aggression directed against outsiders. In human beings, too, the feeling of togetherness which is so essential to the serving of a common cause is greatly enhanced by the presence of a definite, threatening enemy whom it is possible to hate. Also, it is much easier to make people identify with a simple and concrete common cause than with an abstract idea. For all these reasons, the teachers of militant ideologies have an enviably easy job in converting young people. We must face the fact that in Russia as well as in China the younger generation knows perfectly well what it is fighting for, while in our culture it is casting about in vain for causes worth embracing. The way in which huge numbers of young Americans have recently identified themselves with the rights of the American Negro is a glorious exception, though the fervor with which they have done so tends to accentuate the prevalent lack of militant enthusiasm for other equally just and equally important causes—such as the prevention of war in general. The actual warmonger, of course, has the best chances of arousing militant enthusiasm because he can always work his dummy or fiction of an enemy for all it is worth.

In all these respects, the defender of peace is at a decided disadvantage. Everything he lives and works for, all the high goals at which he aims, are, or should be, determined by moral responsibility which presupposes quite a lot of knowledge and real insight. Nobody can get really enthusiastic about them without considerable erudition. The one and only unquestionable value that can be appreciated independently of rational morality or education is the bond of human love and friendship from which all kindness and charity springs, and which represents the great antithesis to aggression. In fact, love and friendship come far nearer to typifying all that is good, than aggression, which is identified with a destructive death drive only by mistake, comes to exemplifying all that is evil.

The champion of peace is debarred from inventing a sort of dummy figure of evil for the purpose of arousing the militant enthusiasm or strengthening the bond between the fighters for the good cause. To attack just "evil" is a questionable procedure, even with intelligent people. Evil, by definition, is that which endangers the good, and the good is that which we perceive as a value. Since for the scientist knowledge represents the highest of all values, he sees the lowest of all negative values in everything that impedes its progress. In my own case, the dangerous whispering of my aggression drive would probably persuade me to see the personification of evil in some philosophers who despise natural science, particularly in those who, for purely ideological reasons, refuse to believe in evolution. If I did not know all that I do about aggression and the compulsion of militant enthusiasm, I should perhaps be in danger of letting myself be inveigled into a religious war against anti-evolutionists. In other words, we had better dispense with the personification of evil, because it leads, all too easily, to the most dangerous kind of war: religious war.

If I have just said that considerable erudition is necessary for anyone to grasp the real values of humanity which are worthy of being served and defended, I certainly did not mean that it was a hopeless task to raise the education of average humanity to that level. I only wanted to emphasize that it was necessary to do so. Indeed, in our age of enlightenment, human beings of average intelligence are not so very far from appreciating real cultural and ethical values. There are at least three great human enterprises, collective in the truest sense of the word, whose ultimate and unconditional value no normal human being can doubt: Art, the pursuit of beauty; Science, the pursuit of truth; and, as an independent third which is neither art nor science, though it makes use of both, Medicine, the attempt to mitigate human suffering.

Not even the most ruthlessly daring demagogues have ever undertaken to proclaim the whole art of an enemy nation or political party as entirely worthless. No normal educated human being can help appreciating the art of another culture however much he finds abhorrent in it in other respects. In addition, painting and music are unhindered by language barriers and are thus able to tell people on one side of a cultural barrier that on its other side, too, there are human beings serving the good and the beautiful. The universal appreciation of Negro music is perhaps an important step toward the solution of the burning racial problem in America. After Negroes had been robbed of their freedom and their own cultural traditions had been successfully extinguished, racial pride and prejudice have done their best, or to be more exact their worst, to prevent them from entering into the spirit and acquiring the basic social norms of Western culture. The only great cultural value which they were not prevented from making wholly their own was music. The indubitable creative power of Negro composers and musicians casts a strong doubt, to say the least, on the alleged lack of cultural creativity of their race.

Art is called upon to create supranational, suprapolitical values that cannot be denied by any narrowly national or political group. It turns traitor to its great mission when it allows itself to be harnessed to any political aim whatsoever. Propagandist tendency in any art, in poetry or in painting, means its final desecration and is altogether evil. Music, though supremely capable of whipping up militant enthusiasm, is fortunately quite unable to specify what the hearers are expected to get enthusiastic about. So the most feudalistic old aristocrat can appreciate the inspiring beauty of the "Marseillaise," even though the text of the song suggests that his impure blood should be used as a fertilizer—*"d'un sang impur abreuvez nos sillons."*

Science, which is closely akin to art in many other respects also, shares its mission of creating a value that no one can deny regardless of his national or political allegiance. Unlike art, science can only be communicated by language, and the truth of its results does not impress as immediately as the beauty of a work of art. On the other hand, opinions concerning the relative value of works of art may differ, and though the true and the false may also be distinguished in art, these words have a very different meaning when applied to the results of scientific research. Truth, in science, can be defined as the working hypothesis best fitted to open the way to the next better one. The scientist knows very well that he is approaching ultimate truth only in an asymptotic curve and is barred from ever reaching it; but at the same time he is proudly aware of being indeed able to determine whether a statement is a nearer or less near approach to truth. This determination is not furnished by any personal opinion nor by the authority of an individual, but by further research proceeding by rules universally accepted by all men of all cultures and all political affiliations. More than any other product of human culture, scientific knowledge is the collective property of all mankind.

Scientific truth is universal, because it is only discovered by the human brain and not made by it, as art is; even philosophy is certainly nothing other than poetry in the original sense of the word, which is derived from the Greek verb *Ποιεὶν*, to make. Scientific truth is wrested from reality existing outside and independent of the human brain. Since this reality is the same for all human beings, all correct scientific results will always agree with each other, in whatever national or political surroundings they may be gained. Should a scientist, in the conscious or even unconscious wish to make his results agree with his political doctrine, falsify or color the results of his work, be it ever so slightly, reality will put in an insuperable veto: these particular results will simply fail on practical application. For example, there was, a few years ago, a school of genetics in the Soviet Union which, from political and, I hope and believe, unconscious reasons, asserted that it had demonstrated the inheritance of acquired characters. These results could not be confirmed anywhere else in the world, and the situation was deeply disturbing to those who believe in the unity of science and its world-embracing mission. There is no more talk of this theory now; geneticists all over the world are again of one opinion. A small victory, indeed, but a victory for truth!

I need not say anything about the general recognition of the value of medicine. The sanctity of the Red Cross is about the only one of the laws of nations that has always been more or less respected by all nations.

Of course, education alone, in the sense of the simple transmission of knowledge, is only a prerequisite to the real appreciation of these and other ethical values. Another condition, quite as important, is that this knowledge and its ethical consequences should be handed down to the younger generation in such a way that it is able to identify itself with these values. I have already said what psychoanalysts have known for a long time, that a relation of trust and respect between two generations must exist in order to make a tradition of values possible. I have already said that Western culture, even without the danger of nuclear warfare, is more directly threatened by disintegration because of its failure to transmit its cultural and even its ethical values to the younger generation. To many people, and probably to all of those actively concerned with politics, my hope of improving the chances of permanent peace by arousing, in young people, militant enthusiasm for the ideals of art, science, medicine and the like, will appear unrealistic to the point of being fatuous. Young people today, they will argue, are notoriously materialistic and take an insuperably skeptical view of ideals in general and in particular of those that arouse the enthusiasm of a member of the older generation. My answer is that this is quite true, but that young people today have excellent excuses for taking this attitude. Cultural and political ideas today have a way of becoming obsolete surprisingly fast; indeed there are few of them on either side of any curtain that have not already done so. To the extraterrestrial observer in whose place we should be trying to put ourselves, it would seem a very minor issue whether capitalism or communism will rule the world, since the differences between the two are rapidly decreasing anyhow. To such an observer, the great questions would be, firstly, whether mankind can keep its planet from becoming too radioactive to support life, and secondly whether mankind will succeed in preventing its population from "exploding" in a way more annihilating than the explosion of the Bomb. Apart from the obvious obsolescence of most so-called ideals, we know some of the reasons why the younger generation refuses to accept handed-down customs and social norms (page 266). I believe that the "angry young men" of Western civilization have a perfectly good right to be angry with the older generation, and I do not regard it as surprising if young people today are skeptical to the point of nihilism. I believe that their mistrust of all ideals is largely due to the fact that there have been and still are so many artificially contrived pseudo-ideals "on the market," calculated to arouse enthusiasm for demagogic purposes.

I believe that among the genuine values here discussed, science has a particular mission in vanquishing this distrust. Honest research must produce identical results anywhere. The verifiability of science proves the honesty of its work. There is no mystery whatsoever about its results; where they are met with obstinate incredulity they can be proved by incontestable figures. I believe that the most materialistic and the most skeptical are the very people whose enthusiasm could be aroused in the service of scientific truth and all that goes with it.

Of course, it is not to be suggested that all of the earth's population should engage in active scientific research, but scientific education might very well become general enough to exert a decisive influence on the social norms approved by public opinion. I am not speaking, at the moment, of the influence which a deeper understanding of the biological laws governing our own behavior might have, a subject I shall discuss later on, but of the beneficial effect of scientific education in general. The discipline of scientific thinking rarely fails to imbue a good man not only with a certain ingrained habit of being honest, but also with a high appreciation of the value of scientific truth in itself. Scientific truth is one of the best causes for which a man can fight and although, being based on irreducible fact, it may seem less inspiring than the beauty of art or some of the older ideals possessing the glamour of myth and romance, it surpasses all others in being incontestable, and absolutely independent of cultural, national, and political allegiances.

Enthusiastic identification with any value that is ethical in the sense that its content will stand the test of Kant's categorical question, will act as an antidote to national or political aggression. Dr. J. Hollo, an American physician, has pointed out that the militant enthusiasm by which a man identifies himself with a national or political cause, is dangerous mainly for the one reason that it excludes all other considerations the moment it is aroused (by the mental processes described on page 268). A man really can feel "wholly American" when thinking of "the" Russians or vice versa. The single-mindedness with which enthusiasm eliminates all other considerations and the fact that the objects of identification happen, in this case, to be fighting units, make national and political enthusiasm actually dangerous, to the point of its being ethically questionable.

Continuing Dr. Hollo's argument, let us suppose that a man, whatever his political or national allegiance, also identifies with ideals other than national or political. Supposing that, being a patriot of my home country (which I am), I felt an unmitigated hostility against another country (which I emphatically do not), I still could not wish wholeheartedly for its destruction if I realized that there were people living in it who, like myself, were enthusiastic workers in the field of inductive natural science, or revered Charles Darwin and were enthusiastically propagating the truth of his discoveries, or still others who shared my appreciation of Michelangelo's art, or my enthusiasm for Goethe's *Faust*, or for the beauty of a coral reef, or for wildlife preservation or a number of minor enthusiasms I could name. I should find it quite impossible to hate, unreservedly, any enemy, if he shared only one of my identifications with cultural and ethical values.

Obviously, the number of cultural and ethical ideals with which people are able to identify irrespective of their national or political allegiance will be in direct proportion to their reluctance to follow the urge of single-minded national or political enthusiasm. It is only the education of all humanity that can increase the number of ideals with which every individual can identify. In this manner, education would become "humanistic" in a new and wider sense of the word.

Humanistic ideals of this kind must become real and full-blooded enough to compete, in the esteem of young people, with all the romantic and glamorous stimulus situations which are, primarily, much more effective in releasing the old hackle-raising and chin-protruding response of militant enthusiasm. Much intelligence and insight, on the side of the educator as well as on that of the educated, will be needed before this great goal is reached. Indeed, a certain academic dryness, unavoidably inherent in humanistic ideals, might forever prevent average humanity from recognizing their value, were it not that they have for their ally a heaven-sent gift of man that is anything but dry, a faculty as specifically human as speech or moral responsibility: humor. In its highest forms, it appears to be specially evolved to give us the power of sifting the true from the false. G. K. Chesterton has voiced the altogether novel opinion that the religion of the future will be based, to a considerable extent, on a more highly developed and differentiated, subtle form of humor. Though, in this formulation, the idea may appear somewhat exaggerated, I feel inclined to agree, answering one paradox with another by saying that we do not as yet take humor seriously enough. I should not write my avowal of optimism with so much conviction were it not for my confidence in the great and beneficial force of humor.

Laughter is not only the overt expression of humor, but it very probably constitutes the phylogenetic base on which it evolved. Laughter resembles militant enthusiasm as well as the triumph ceremony of geese in three essential points: all three are instinctive behavior patterns, all three are derived from aggressive behavior and still retain some of its primal motivation, and all three have a similar social function. As discussed in Chapter Five, laughter probably evolved by ritualization of a redirected threatening movement, just as the triumph ceremony did. Like the latter, and like militant enthusiasm, laughter produces, simultaneously, a strong fellow feeling among participants and joint aggressiveness against outsiders. Heartily laughing together at the same thing forms an immediate bond, much as enthusiasm for the same ideal does. Finding the same thing funny is not only a prerequisite to a real friendship, but very often the first step to its formation. Laughter forms a bond and simultaneously draws a line. If you cannot laugh with the others, you feel an outsider, even if the laughter is in no way directed against yourself or indeed against anything at all. If laughter is in fact directed at an outsider, as in scornful derision, the component of aggressive motivation and, at the same time, the analogy to certain forms of the triumph ceremony become greatly enhanced. In this case, laughter can turn into a very cruel weapon, causing injury if it strikes a defenseless human being undeservedly: it is criminal to laugh at a child.

Nevertheless laughter is, in a higher sense than enthusiasm, specifically human. The motor patterns of threatening underlying both have undergone a deeper change of form and function in the case of laughter. Unlike enthusiasm, laughter—even at its most intense—is never in danger of regressing and causing the primal aggressive behavior to break through. Barking dogs may occasionally bite, but laughing men hardly ever shoot! And if the motor patterns of laughing are even more uncontrollably instinctive than those of enthusiasm, conversely its releasing mechanisms are far better and more reliably controlled by human reason. Laughter never makes us uncritical, while enthusiasm abolishes all thought of rational self-control.

Indeed, the reliable control exerted by reason over laughter allows us to use it in a way which would be highly dangerous if applied to militant enthusiasm. Both laughter and enthusiasm can, by appropriate manipulation, be used like aggressive dogs, that is, set on and made to attack practically any enemy that reason may choose. But while laughter, even in the form of the most outrageous and scornful ridicule, always remains obedient to reason, enthusiasm is always threatening to get out of hand and to turn on its master.

There is one particular enemy whom it is fair to attack to the barking tune of laughter, and that is a very definite form of lie. There are few things in the world so thoroughly despicable and deserving of immediate destruction as the fiction of an ideal cause artificially set up to elicit enthusiasm in the service of the contriver's aims. Humor is the best of lie-detectors, and it discovers, with an uncanny flair, the speciousness of contrived ideals and the insincerity of simulated enthusiasm. There are few things in the world so irresistibly comical as the sudden unmasking of this sort of pretense. When pompousness is abruptly debunked, when the balloon of puffed-up arrogance is pricked by humor and bursts with a loud report, we can indulge in uninhibited, refreshing laughter which is liberated by this special kind of sudden relief of tension. It is one of the few absolutely uncontrolled discharges of an instinctive motor pattern in man of which responsible morality wholly approves.

Responsible morality not only approves of the effects of humor, but finds a strong support in it. A satire is, by the definition of the Concise Oxford Dictionary, a poem aimed at prevalent vices and follies. The power of its persuasion lies in the manner of its appeal: it can make itself heard by ears which, through skepticism and sophistication, are deafened to any direct preaching of morality. In other words, satire is the right sort of sermon for today.

If, in ridiculing insincere ideals, humor is a powerful ally of rational morality, it is even more so in self-ridicule. Nowadays we are all radically intolerant of pompous or sanctimonious people, because we expect a certain amount of self-ridicule in every intelligent human being. Indeed, we feel that a man who takes himself absolutely seriously is not quite human, and this feeling has a sound foundation. That which, in colloquial German, is so aptly termed "*tierischer Ernst*"—that is, "animal seriousness"—is an ever present symptom of megalomania, in fact I suspect that it is one of its causes. The best definition of man is that he is the one creature capable of reflection, of seeing himself in the frame of reference of the surrounding universe. Pride is one of the chief obstacles to seeing ourselves as we really are, and self-deceit is the obliging servant of pride. It is my firm belief that a man sufficiently gifted with humor is in small danger of succumbing to flattering delusions about himself because he cannot help perceiving what a pompous ass he would become if he did. I believe that a really subtle and acute perception of the humorous aspects of ourselves is the strongest inducement in the world to make us honest with ourselves, thus fulfilling one of the first postulates of reasoning morality. An amazing parallel between humor and the categorical question is that both balk at logical inconsistencies and incongruities. Acting against reason is not only immoral but, funnily enough, it is very often extremely funny! "Thou shalt not cheat thyself" ought to be the first of all commandments. The ability to obey it is in direct proportion to the ability of being honest with others.

It is not only because of these considerations that I regard humor as a force which justifies greater optimism. I also believe that humor is rapidly developing in modern man. Whether humor is becoming more effective because cultural tradition makes it more and more respected, or whether the instinctive drive to laugh is phylogenetically gaining power, is not the essential point; both processes probably are at work. In any case, there is no doubt that humor is rapidly becoming more effective, more searching, and more subtle in detecting dishonesty. I for one find the humor of earlier periods less effective, less probing, less subtle. Charles Dickens is the oldest writer I know whose satirical representation of human nature makes me really laugh. I can understand perfectly well at what particular "prevalent vices and follies" the satires of late Roman writers or of Abraham a Sancta Clara are directed, but I do not respond to them with laughter. A systematic historical investigation of the stimulus situations that in different ages made people laugh might be extremely revealing.

I believe that humor exerts an influence on the social behavior of man which, in one respect, is strictly analogous to that of moral responsibility: it tends to make the world a more honest and, therewith, a better place. I believe that this influence is rapidly increasing and, entering more and more subtly into the reasoning processes, becoming more closely interwoven with and, in its effects, more akin to morality. In this sense, I absolutely agree with G. K. Chesterton's astonishing statement.

From the discussion of what I know I have gradually passed to the account of what I think probable and, finally, to a profession of what I believe. There is no law barring the scientist from doing so. I believe, in short, in the ultimate victory of truth. I know that this sounds rather pompous, but I honestly do think it is the most likely thing to happen. I might even say that I regard it as inevitable, provided the human species does not commit suicide in the near future, as well it may. Otherwise it is quite predictable that the simple truths concerning the biology of mankind and the laws governing its behavior will sooner or later become generally accepted public property, in the same way as the older scientific truths discussed in Chapter Twelve have done; they, too, were at first unacceptable to an all too complacent humanity because they disturbed its exaggerated self-esteem. Is it too much to hope that the fear of imminent self-destruction may have a sobering effect and act as a monitor of self-knowledge?

I do not consider in any way as utopian the possibility of conveying a sufficient knowledge of the essential biological facts to any sensible human being. They are indeed much easier to understand than, for instance, integral calculus or the computing of compound interest. Moreover, biology is a fascinating study, provided it is taught intelligently enough to make the pupil realize that he himself, being a living being, is directly concerned with what he is being told. "*Tua res agi-*

tur." Expert teaching of biology is the one and only foundation on which really sound opinions about mankind and its relation to the universe can be built. Philosophical anthropology of a type neglecting biological fact has done its worst by imbuing humanity with that sort of pride which not only comes before, but causes, a fall. It is plain biology of Homo sapiens L. that ought to be considered the "big science."

Sufficient knowledge of man and of his position in the universe would, as I have said, automatically determine the ideals for which we have to strive. Sufficient humor may make mankind blessedly intolerant of phony, fraudulent ideals. Humor and knowledge are the two great hopes of civilization. There is a third, more distant hope based on the possibilities of human evolution; it is to be hoped that the cultural factors just mentioned will exert a selection pressure in a desirable direction. Many characters of man which, from the Paleolithic to recent times, were accounted the highest virtues, today seem dangerous to thinking people and funny to people with a sense of humor. If it is true that, within a few hundred years, selection brought about a devastating hypertrophy of aggression in the Utes, that most unhappy of all peoples, we may hope without exaggerated optimism that a new kind of selection may, in civilized peoples, reduce the aggressive drive to a tolerable measure without, however, disturbing its indispensable function.

The great constructors of evolution will solve the problems of political strife and warfare, but they will not do so by entirely eliminating aggression and its communal form of militant enthusiasm. This would not be in keeping with their proven methods. If, in a newly arising biological situation, a drive begins to become injurious, it is never atrophied and removed entirely, for this would mean dispensing with all its indispensable functions. Invariably, the problem is solved by the evolution of a new inhibitory mechanism adapted to dealing specifically with the new situation and obviating the particular detrimental effects of the drive without otherwise interfering with its functions.

We know that, in the evolution of vertebrates, the bond of personal love and friendship was the epoch-making invention created by the great constructors when it became necessary for two or more individuals of an aggressive species to live peacefully together and to work for a common end. We know that human society is built on the foundation of this bond, but we have to recognize the fact that the bond has become too limited to encompass all that it should: it prevents aggression only between those who know each other and are friends, while obviously it is all active hostility between all men of all nations or ideologies that must be stopped. The obvious conclusion is that love and friendship should embrace all humanity, that we should love all our human brothers indiscriminately. This commandment is not new. Our reason is quite able to understand its necessity as our feeling is able to appreciate its beauty, but nevertheless, made as we are, we are unable to obey it. We can feel the full, warm emotion of friendship and love only for individuals, and the utmost exertion of will power cannot alter this fact. But the great constructors can, and I believe they will. I believe in the power of human reason, as I believe in the power of natural selection. I believe that reason can and will exert a selection pressure in the right direction. I believe that this, in the not too distant future, will endow our descendants with the faculty of fulfilling the greatest and most beautiful of all commandments.

Bibliography

ALTMANN, M. "A Study of Behavior in a Horse-Mule Group." *Sociometry*, 14 (1951), 351–354.

—— "Patterns of Social Behavior in Big Game of the United States and Europe." *Trans. North Amer. Wildl. Conf.*, 21 (1956), 538–545.

—— "Group Dynamics in Wyoming Moose During the Rutting Season." *Journal of Mammalogy*, 40 (1959), 420–424.

ALTMANN, S. A. "A Field Study of the Sociobiology of Rhesus Monkeys, Macaca mulatta." *Ann. N. Y. Acad. Sci.*, 102 (1962), 338–435.

ANDREW, R. J. "The Origin and Evolution of the Calls and Facial Expressions of the Primates." *Behavior*, 20 (1963), 1–109.

BAERENDS, G. P., and J. M. BAERENDS VAN ROON. "An Introduction to the Study of the Ethology of Cichlid Fishes." *Behavior Suppl.* 1 (1950).

BAEUMER, E. "Das dumme Huhn: Verhalten des Haushuhns." *Kosmos-Bibliothek*, 242 (1964).

BOWLBY, J. "Maternal Care and Mental Health." World Health Organisation, Monogr. Ser. No. 2 (1952).

CARPENTER, C. R. "A Field Study of the Behavior and Social Relations of Howling Monkeys." *Comp. Psychol. Monogr.*, 10 (1934), 1–168.

CRAIG, W. "Appetites and Aversions as Constituents of Instincts." *Biol. Bull.*, 34 (1918), 91–107.

DARLING, F. *A Herd of Red Deer*. New York: Oxford University Press, 1937.

DARWIN, C. *Origin of Species*. London: John Murray, 1859.

—— *The Expression of the Emotions in Man and Animals*. London: John Murray, 1872.

DeVORE, I. *Primate Behavior: Field Studies of Monkeys and Apes.* New York: Holt, Rinehart and Winston, 1965.

EIBL-EIBESFELDT, I. "Beiträge zur Biologie der Haus- und Ährenmaus nebst einigen Beobachtungen an anderen Nagern." *Z. Tierpsychol.,* 7 (1950), 558–587.

——— "Ethologische Unterschiede zwischen Hausund Wanderratte." *Verh. Dtsch. Zool. Ges.* Freiburg (1952), 169–180.

——— "Zur Ethologie des Hamsters (Cricetus cricetus L.)." *Z. Tierpsychol.,* 10 (1953), 204–254.

ERIKSON, E. H. "Ontogeny of Ritualisation in Man." *Philosophical Transactions,* Royal Society, London 251 B (1966), 337–349.

FISCHER, H. "Das Triumphgeschrei der Graugans (Anser anser)." *Ibid.,* 22 (1965), 247–304.

GEHLEN, A. *Der Mensch, seine Natur und seine Stellung in der Welt.* Berlin: Junker und Dürrhaupt, 1940.

HASSENSTEIN, B. "Abbildende Begriffe." *Ver. Dtsch. Zool. Ges.* Tübingen (1954), 197–202.

HEDIGER, H. "Zur Biologie und Psychologie der Flucht bei Tieren." *Biol. Zbl.,* 54 (1934), 21–40.

——— "Säugetiersoziologie." *Colloques Internationaux du Centre National de la Recherche Scientifique,* 34 (1950), 297–321.

——— *Wild Animals in Captivity: An Outline of the Biology of Zoological Gardens.* New York: Dover, 1965.

——— *Studies of the Psychology and Behaviour of Captive Animals in Zoos and Circuses.* London: 1955.

HEILIGENBERG, W. "Ursachen für das Auftreten von Instinktbewegungen bei einem Fische (Pelmatochromis subocellatus kribensis Boul. Cichlidae)." *Zur vergl. Physiol.,* 47 (1963), 339–380.

HEINROTH, O. "Beiträge zur Biologie, insbesondere Psychologie und Ethologie der Anatiden." *Verh. d. 5. Intern. Ornithol. Kongr. Berlin* (1910).

——— and M. HEINROTH. *Die Vögel Mitteleuropas.* Berlin: Bermühler, 1924–28.

HINDE, R. A. "Ethological Models and the Concept of 'Drive.'" *Brit. J. Phil. Sci.,* 6 (1956), 321–331.

——— "Unitary Drives." *Anim. Behaviour,* 7 (1959), 130–141.

——— "Factors Governing the Changes in Strength of a Partially Inborn Response, as Shown by the Mobbing Behaviour of the Chaffinch (Fringilla coelebs)." *Proc. Roy. Soc. B.,* 753 (1960), 398–420.

v. HOLST, E. "Über den Prozess der zentralnervösen Koordination." *Pflüg. Arch.*, 236 (1935), 149–158.

—— "Alles oder Nichts, Block, Alternans, Bigemini und verwandte Phänomene als Eigenschaften des Rückenmarks." *Ibid.*, 236 (1935), 4, 5, 6.

—— "Versuche zur Theorie der relativen Koordination." *Ibid.*, 237 (1936).

—— "Vom Dualismus der motorischen und der automatisch-rhythmischen Funktion im Rückenmark und vom Wesen des automatischen Rhythmus." *Ibid.*, 237 (1936), 3.

—— "Regelvorgänge in der optischen Wahrnehmung." *Rept. 5th Conf. Soc. Biol. Rhythmn.* Stockholm, 1955.

HUXLEY, J. S. "The Courtship-Habits of the Great Crested Grebe (Podiceps cristatus); with an Addition to the Theory of Sexual Selection." *Proc. Zool. Soc. London,* 35 (1914), 491–562.

—— *Evolution, the Modern Synthesis.* New York: Harper, 1942.

—— *Evolution in Action.* New York: The New American Library, 1957.

—— *Essays of a Humanist.* New York: Harper, 1964.

—— and J. Huxley. *Evolution and Ethics.* London: The Pilot Press, 1947.

IERSEL, J. van. "An Analysis of the Parental Behaviour of the Breed-spined Stickleback (Gasterosteus aculeatus)." *Beh. Suppl.* 3 (1953).

KIRCHSHOFER, R. "Aktionssystem des Maulbrüters Haplochromis desfontainesii." *Z. Tierpsychol.*, 10 (1953), 297–318.

KITZLER, G. "Die Paarungsbiologie einiger Eidechsen." *Ibid.*, 4 (1942), 353.

KOEHLER, O. "Die Ganzheitsbetrachtung in der modernen Biologie." *Verhandl. d. Königsberger gelehrten Ges.* (1933).

LEYHAUSEN, P. "Die Entdeckung der relativen Koordination: Ein Beitrag zur Annäherung von Physiologie und Psychologie." *Studium Generale,* 7 (1954), 45–60.

—— "Über relative Stimmungshierarchie bei Säugern." Lecture at the 3rd International Conference of Ethologists, Groningen. (1955.) Unpublished.

—— "Verhaltensstudien an Katzen." *Z. Tierpsychol.* (1956). Suppl. 2.

—— "Über den Begriff des Normalverhaltens in der Ethologie." Lecture at the 3rd International Conference of Ethologists, Starnberg. (1961.) Unpublished.

LEYHAUSEN, P. "Über die Funktion der relativen Stimmungshierarchie. (Dargestellt am Beispiel der phylogenetischen und ontogenetischen Entwicklung des Beutefangs von Raubtieren)." *Z. Tierpsychol.*, 22 (1965), 412–494.

—— "Das Motivationsproblem in der Ethologie." *Hdbch. Psychol. Bd. Motivationslehre,* Göttingen: 1965.

—— and R. Wolff. "Das Revier einer Hauskatze." *Z. Tierpsychol.*, 16 (1959), 66–70.

LORENZ, K. "Beobachtungen an Dohlen." *J. Ornithol.*, 75 (1927), 511–519.

—— "Beiträge zur Ethologie sozialer Corviden." *Ibid.*, 79 (1931), 67–127.

—— "Beobachtungen an freifliegenden zahmgehaltenen Nachtreihern." *Ibid.*, 82 (1934), 160–161.

—— "Contribution to the Comparative Sociology of Colony-nesting Birds." *Proc. VIII. Internat. Ornithol. Congr.* (1934), 207–218.

—— "Der Kumpan in der Umwelt des Vogels." *J. Ornithol.*, 83 (1935), 137–215, 289–413.

—— "Über die Bildung des Instinktbegriffes." *Naturwiss.*, 25 (1937), 298–300, 307–308, 324–331.

—— "Vergleichende Verhaltensforschung." *Verh. Dtsch. Zool. Anz. Suppl.*, 12 (1939), 69–102.

—— "Die Paarbildung beim Kolkraben." *Z. Tierpsychol.*, 3 (1940), 278–292.

—— "Vergleichende Bewegungsstudien an Anatiden." *J. Ornithol.*, 89 (1941), Suppl. 3, Festschrift O. Heinroth, 194–293.

—— "Ganzheit und Teil in der tierischen und menschlichen Gemeinschaft." *Studium Generale,* 9 (1950), 455–499.

—— "The Comparative Method in Studying Innate Behavior Patterns." *Symp. Soc. Exp. Biol.*, 4 (1950), Animal Behaviour, Cambridge, 221–268.

—— "Die Entwicklung der vergleichenden Verhaltensforschung in den letzten 12 Jahren." *Verh. Dtsch. Zool. Ges.* Freiburg (1952), 36–58.

—— "Psychologie und Stammesgeschichte." In G. Heberer, ed.: *Die Evolution der Organismen.* Jena: G. Fischer, 1954, 2nd ed., 131–172.

—— "Über das Töten von Artgenossen." *Jahrb. d. Max-Planck-Ges.* (1955), 105–140.

—— "Moralanaloges Verhalten geselliger Tiere." *Universitas,* 11/7 (1956), 691–704.

—— "The Role of Aggression in Group Formation." *Transactions of the 4th Conference on Group Processes*. New York: Josiah Macy Jr. Foundation, 1957.

—— *Methods of Approach to the Problems of Behavior*. The Harvey Lectures, Ser. 54 (1959), 60–103. New York, Academic Press.

—— "Prinzipien der vergleichenden Verhaltensforschung." *Fortschr. Zool.*, 12 (1960), 265–294.

—— "Gestalt Perception as Fundamental to Scientific Knowledge." General Systems, New York. 7 (1962), 37–56.

—— "The Function of Colour in Coral Reef Fishes." *Proc. Roy. Inst. of Great Britain*, 39 (1962), 282–296.

McDOUGALL, W. "Outline of Psychology." Methuen & Co., London (1933).

MARGOLIN, S. "A Consideration of Constitutional Factors in Aggressivity of an Indian Tribe," lecture delivered at the Menninger School of Psychiatry, Topeka, 1960.

MASSERMANN, J. H. *Behavior and Neurosis*. Chicago: University of Chicago Press, 1943.

MEYER-HOLZAPFEL, M. "Triebbedingte Ruhezustände als Ziel von Appetenzhandlüngen." *Naturwiss.*, 28 (1940), 273–280.

MORRIS, D. "Typical Intensity and Its Relation to the Problem of Ritualisation." *Behaviour* 11 (1957), 1–12.

MURIE, A. *The Wolves of Mount McKinley*. Fauna of the National Parks of the United States Series 5. 1944.

NICOLAI, J. "Zur Biologie und Ethologie des Gimpels." *Z. Tierpsychol.*, 13 (1956), 93–132.

OEHLERT, B. "Kampf und Paarbildung einiger Cichliden." *Ibid.*, 15 (1958), 141–174.

PRUITT, W. O. "The Function of the Brow-Tine in Caribou Antlers." *J. Arctic Inst. North America* 19, 2 (1966).

RUSSELL, W. N. "Evolutionary Concepts in Behavioural Science." I–III, General Systems (Year Book of the Society for General Systems Research) 3 (1958), 18–28; 4 (1959), 45–73; 6 (1961), 51–92.

SCHENKEL, R. "Ausdrucksstudien an Wölfen." *Behaviour*, 1 (1947), 81–129.

SCHJELDERUP-EBBE, Th. "Zur Sozialpsychologie der Vögel." *J. Psychol.* 88 (1922), 224.

SCHLEIDT, W. "Über die Spontaneität von Erbkoordinationen." *Z. Tierpsychol.*, 21 (1963), 235–256.

SCHLEIDT, W., and M. Schleidt. "Störung der Mutter-Kind-Beziehung bei Truthühnern durch Gehörverlust." *Behaviour,* 16 (1960), 3-4.

SCHÜZ, E. "Bewegungsnormen des Weissen Storches." *Z. Tierpsychol.,* 5 (1941), 1-38.

SEITZ, A. "Die Paarbildung bei einigen Cichliden. I. Die Paarbildung bei Astatotilapia strigigena." *Ibid.,* 4 (1940), 40-84.

—— "Die Paarbildung bei Hemichromis bimaculatus Gill." *Ibid.,* 5 (1942), 74-101.

SPITZ, R. A. *La première Année de la Vie de l'Enfant.* Paris: Presses Universitaires de France, 1958.

STEINIGER, F. "Zur Soziologie und sonstigen Biologie der Wanderratte." *Z. Tierpsychol.,* 7 (1950), 356-379.

THORPE, W. H. *Learning and Instinct in Animals.* London: Methuen, 1956.

TINBERGEN, N. *Eskimoland.* Rotterdam: 1934.

—— "Die Übersprungbewegung." *Z. Tierpsychol.,* 4 (1940), 1-40.

—— "An Objectivistic Study of the Innate Behaviour of Animals." *Biblioth. biotheor.,* 1 (1942), 39-98.

—— *Instinktlehre.* Berlin: Paul Parey, 1952.

—— *Social Behaviour in Animals.* New York: John Wiley, 1953.

—— "Some Aspects of Ethology, the Biological Study of Animal Behaviour." *Advan. Sci.,* 12 (1955), 17-27.

—— *The Herring Gull's World.* Rev. ed. New York: Basic Books, 1961.

—— "On Aims and Methods of Ethology." *Z. Tierpsychol.,* 20 (1963), 404-433.

WASHBURN, S. L., and I. DeVORE. "The Social Life of Baboons." *Sci. Am.,* 204/6 (1961), 62-71.

WHITMAN, C. O. "The Behavior of Pigeons." Carnegie Inst. Wash. Publ. 257 (1919), 1-161.

WIEPKEMA, P. R. "An Ethological Analysis of the Reproductive Behaviour of the Bitterling (Rodeus Amarus Block)." *Arch. néerl. Zool.,* 14 (1961), 103-199.

YERKES, R. M. *Chimpanzees: A Laboratory Colony.* New Haven: Yale University Press, 1943.

ZUMPE, D. "Laboratory Observations on the Aggressive Behaviour of Some Butterfly Fishes (Chaetodontidae)." *Z. Tierpsychol.,* 22 (1965), 226-236.